Practical physiology

(Laboratory methods and techniques)

I0503447

Ayman Mohamed

ISBN: 1515352714
ISBN-13: 978-1515352716

Ayman Mohamed

DEDICATION

I dedicate my book to my family and many friends. A special feeling of gratitude to my loving parents, Saber Mohamed and Karema Ramadan, whose words of encouragement and push for tenacity ring in my ears. I dedicate this work and give special thanks to my aunt Dr. Sohair R. Fahmi

CONTENTS

ACKNOWLEDGMENTS

Firstly, I thank Allah for his grace, his success and his generosity.
I take this opportunity to express my profound gratitude and deep regards to Dr. Sohair R. Fahmi, Associate Professor of Physiology, Zoology Department, Faculty of Science, Cairo University, for supervising the present work, his guidance, monitoring and constant encouragement during the work.
I would like to express my sincere gratitude to Dr. Amel Mahmoud Ali Soliman, Associate Professor of Physiology, Zoology Department, Faculty of Science, Cairo University, for her supervision, valuable advices, stimulating suggestions through this work.
Special thanks to Prof. Dr. Mohamed Assem Said Marie, Professor of Environmental Physiology, Zoology Department, Faculty of Science, Cairo University, for his help and support.
I deeply thank my family for their love, support and encouragement through the work.

CHAPTER 1
PHYSIOLOGICAL BUFFER

I.   Buffer definition
A buffer solution is one which maintains its pH fairly constant even upon the addition of small amounts of acid or base. In other words, a buffer solution resists a change in its pH.

II.   Important of buffer
The maintaining a stable pH when studying enzymes activities is necessary. Biochemical processes can be affected by minute changes in pH. It has become very important to find buffers to stabilize hydrogen ion concentrations while not interfering with the function of the enzyme being studied.

III.   Types of buffer solutions
1- Acid buffers
A weak acid together with its salt of strong base.
Example: $CH_3COOH + CH_3COONa$.

2- Basic buffers
A weak base and its salt with a strong acid.
Example: $NH_4OH + NH_4Cl$.

IV.   Mechanism of buffer action

1- Addition of acid:
Upon the addition of HCl, the increase of H^+ ions is counteracted by association with the excess of acetate ions to form unionized CH_3COOH. Thus the added H^+ ions are neutralized and the pH of the buffer solution remains virtually unchanged.

2- Addition of base
When NaOH is added to the buffer solution, the additional OH^- ions combine with H^+ ions of the buffer to form water molecules.

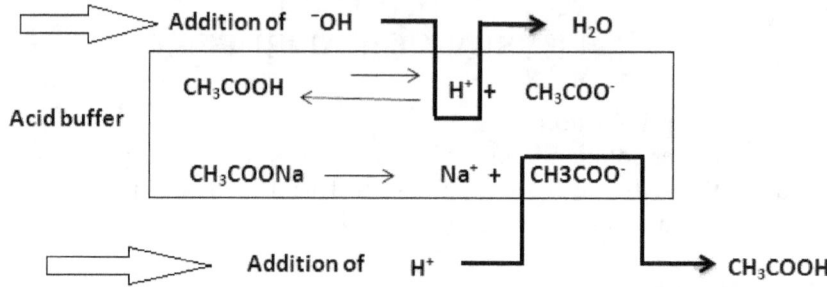

Mechanism of acid buffer action

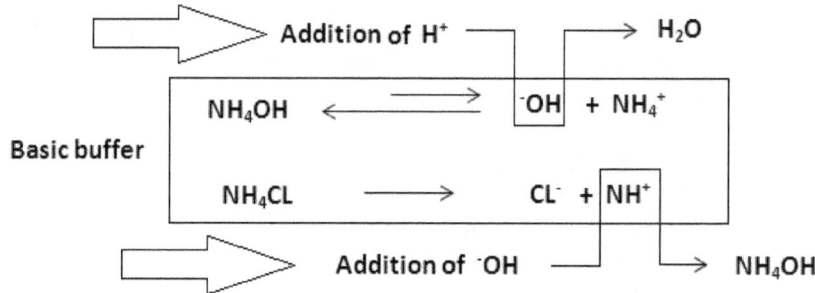

Mechanism of acid buffer action

V. **Criteria of physiological Buffers**

Biological buffers should meet the following general criteria:
1. Their pKa should reside between 6.0 to 8.0.
2. They should exhibit high water solubility and minimal solubility in organic solvents.
3. They should not permeate cell membranes.
4. They should not exhibit any toxicity towards cells.
5. They should not interfere with any biological process.
6. The salt effect should be minimal; however, salts can be added as required.
7. Ionic composition of the medium and temperature should have minimal effect of buffering capacity.
8. Buffers should be stable and resistant to enzymatic degradation.
9. Buffer should not absorb either in the visible or in the UV region.
10. Most of the buffers used in cell cultures, isolation of cells, enzyme assays, and other biological applications must possess

these distinctive characteristics.

VI.   Preparation of Some Common Buffers

The information provided below is intended only as a general guideline. You must use of sensitive pH meter with appropriate temperature setting for final pH adjustment. Addition of other chemicals, after adjusting the pH, may change the final pH value to some extent.

1- Tris-hydrochloride buffer pH range 7.2 to 9.0

Tris is an organic compound with the formula $NH_2C(CH_2OH)_3$

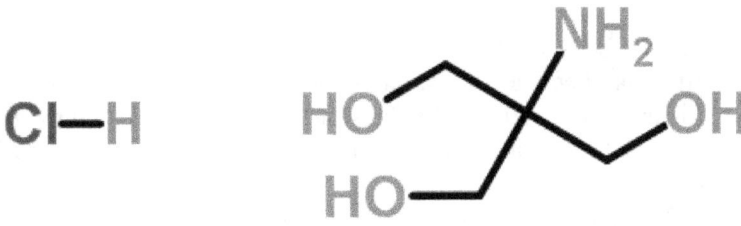

Tris-HCl is excellent biochemical and biological buffer for all applications where ultra-high purity is required. The pH values of all buffers are temperature and concentration dependent. For precise applications, use a carefully calibrated pH meter with a glass/calomel combination electrode

2- Preparation of 1 M Tris solutions (1 liter)
- 121.14 g Tris base in 800 ml dH_2O
- Set pH with concentrated hydrochloric acid (1N)
- dH_2O ad 1000ml

3- Preparation of 50 mM Tris solution (1 liter)
- 6.057 g Tris base in 800 ml liter dH2O
- Set pH with concentrated hydrochloric acid (1N)
- dH_2O ad 1000ml

pH	1 M Tris solution	50 mM Tris solution
	Volume (ml) conc. HCl	
7.2	76.10	44.7
7.5	69.10	40.3
8.0	48.30	29.2
8.5	23.90	14.7
9.0	8.25	7.0

4- Tris-EDTA buffer

This buffer has become the standard buffer for the storage of nucleic acids. It is used at different pH values. It is generally prepared by mixing Tris buffer stock solutions (1M) with an EDTA stock solution (0.5 M; pH 8.0).

- 10 mM Tris- HCl (pH 7.4, 7.5 or 8.0)
- 1 mM EDTA (pH 8.0)

The prepared buffer can be stored at room temperature.

5- Phosphate Buffer; pH range 5.8 to 8.0

The principle of phosphate buffering is independent of the source of phosphates. All that matters for buffering is the ratio of the H2PO4 and HPO4 ions in solution.

a) Phosphate buffer (formula 1)

Stock solution A:

- Potassium dihydrogen phosphate 0.1 M

(Dissolve 13.61 g KH_2PO_4 in 1000 mL distilled water)

Stock solution B

- Disodium hydrogen phosphate 0.1 M

(Dissolve 17.8 g $Na_2HPO_4.2\ H_2O$ in 1000 mL distilled water)

Preparation of the buffer solution: mix stock A and stock B according to the table below:

pH-value	Stock solution B (mL)	Stock solution A (mL)
5.30	2.5	97.5
5.60	5.0	95.0
5.91	10.0	90.0
6.24	20.0	80.0
6.47	30.0	70.0
6.64	40.0	60.0
6.81	50.0	50.0
6.98	60.0	40.0
7.17	70.0	30.0
7.38	80.0	20.0
7.73	90.0	10.0
8.04	95.0	5.0

b) **Phosphate buffer (0.2 M) (formula 2)**
Stock solution A:
- 0.2M monobasic sodium phosphate
27.8 g $NaH_2PO_4.H_2O$ in 1000 ml
Stock solution B:
- 0.2 M dibasic sodium phosphate
53.65 g of Na_2HPO_4 . $7H_2O$ in 1000 ml **or**
71.7 g of Na_2HPO4. $12H_2O$ in 1000 ml
Add (x ml of A, y ml of B, diluted to a total of 200 ml:

pH	Stock solution A	Stock solution B	pH	Stock solution A	Stock solution B
5.7	93.5	6.5	6.9	45.0	55.0
5.8	92.0	8.0	7.0	39.0	61.0
5.9	90.0	10.0	7.1	33.0	67.0
6.0	87.7	12.3	7.2	28.0	72.0
6.1	85.0	15.0	7.3	23.0	77.0
6.2	81.5	18.5	7.4	19.0	81.0
6.3	77.5	22.5	7.5	16.0	84.0
6.4	73.5	26.5	7.6	13.0	87.0
6.5	68.5	31.5	7.7	10.5	90.5
6.6	62.5	37.5	7.8	8.5	91.5
6.7	56.5	43.5	7.9	7.0	93.0
6.8	51.0	49.0	8.0	5.3	94.7

c) Phosphate buffer (0.2 M) (formula 3)

- 50 mL 0f 0.2 M KH_2PO_4 (27.2g)
- Add the indicated number of mL of 0.2 M NaOH (40g)
- Dilute to 200 mL.

pH	mL NaOH	pH	mL NaOH
5.8	3.72	7.0	29.63
6.0	5.70	7.2	35.00
6.2	8.60	7.4	39.50
6.4	12.60	7.6	42.80
6.6	17.80	7.8	45.20
6.8	23.65	8.0	46.80

6- Phosphate buffered saline systems (PBS)

Laboratories may follow several ways for the preparation of phosphate buffered saline solutions.

a) Phosphate buffer 0.01 M pH 7.4 containing 0.9% NaCl

- Disodium hydrogen phosphate ($Na_2HPO_4.12 H_2O$) = 2.76g
- Sodium dihydrogen phosphate ($NaH_2PO_4 .2 H_2O$) = 0.35 g
- Sodium chloride (NaCl) = 9.0 g
- Dissolve in 1000 ml distilled water and adjust pH to 7.4

b) Phosphate buffered saline (10x PBS) 0.1 M pH 7.2

- Disodium hydrogen phosphate (Na_2HPO_4 anhydrous) = 10.9 g
- Sodium dihydrogen phosphate (NaH_2PO_4 anhydrous) = 3.2g
- Sodium chloride (NaCl) = 90.0 g
- Dissolve in 1000 ml distilled water and adjust pH to 7.2
- Dilute 1:10 with distilled water prior to use and adjust pH if necessary

c) 1X Phosphate Buffered Saline

- Dissolve the following in 800ml distilled H2O.
  - 8g of NaCl
  - 0.2g of KCl
  - 1.44g of Na2HPO4
  - 0.24g of KH2PO4

- Adjust pH to 7.4.
- Adjust volume to 1L with additional distilled H2O.

7- Citrate buffer stock solutions

- A: 0.1 M solution of citric acid (21.01 g $C_6H_8O_7.H_2O$ in 1000 ml)
- B: 0.1 M solution of sodium citrate (29.41 g $C_6H_5O_7Na_3$. $2H_2O$ in 1000 ml)
- x ml of A and y ml of B, diluted to a total of 100 ml:

X	y	pH	X	y	pH
46.5	3.5	3.0	23.0	27.0	4.8
43.7	6.3	3.2	20.5	29.5	5.0
40.0	10.0	3.4	18.0	32.0	5.2
37.0	13.0	3.6	16.0	34.0	5.4
35.0	15.0	3.8	13.7	36.3	5.6
33.0	17.0	4.0	11.8	38.2	5.8
31.5	18.5	4.2	9.5	41.5	6.0
28.0	22.0	4.4	7.2	42.8	6.2
25.5	24.5	4.6			

8- Citrate–phosphate buffer stock solutions

- A: 0.1 M solution of citric acid (19.21 g in 1000 ml)
- B: 0.2 M solution of dibasic sodium phosphate (53.65 g of Na_2HPO_4. $7H_2O$ or 71.7 g of $Na_2HPO_4.12H_2O$ in 1000 ml)
- x ml of Ay ml of B, diluted to a total of 100 ml:

x	y	pH	x	y	pH
44.6	5.4	2.6	24.3	25.7	5.0
42.2	7.8	2.8	23.3	26.7	5.2
39.8	10.2	3.0	22.2	27.8	5.4
37.7	12.3	3.2	21.0	29.0	5.6
35.9	14.1	3.4	19.7	30.3	5.8
33.9	16.1	3.6	17.9	32.1	6.0
32.3	17.7	3.8	16.9	33.1	6.2
30.7	19.3	4.0	15.4	34.6	6.4
29.4	20.6	4.2	13.6	36.4	6.6
27.8	22.2	4.4	9.1	40.9	6.8
26.7	23.3	4.6	6.5	43.6	7.0
25.2	24.8	4.8	24.3		

9- Acetate buffer stock solutions
- A: 0.2 M solution of acetic acid CH_3CO_2H (11.55 ml in 1000 ml)
- B: 0.2 M solution of sodium acetate (16.4 g of $C_2H_3O_2Na$ or 27.2 g of $C_2H_3O_2Na$. $3H_2O$ in 1000 ml)
- x ml of Ay ml of B, diluted to a total of 100 ml:

x	y	pH
46.3	3.7	3.6
44.0	6.0	3.8
41.0	9.0	4.0
36.8	13.2	4.2
30.5	19.5	4.4
25.5	24.5	4.6
14.8	35.2	5.0
10.5	39.5	5.2
8.8	41.2	5.4
4.8	45.2	5.6

10- Cacodylate buffer stock solutions

- A: 0.2 M solution of sodium cacodylate (42.8 g of $Na(CH_3)2AsO_2.3H_2O$ in 1000 ml)
- B: 0.2 M NaOH
- 50 ml of A, x ml of B, diluted to a total of 200 ml:

X	pH	x	pH
2.7	7.4	29.6	6.0
4.2	7.2	34.8	5.8
6.3	7.0	39.2	5.6
9.3	6.8	43.0	5.4
13.3	6.6	45.0	5.2
18.3	6.4	47.0	5.0
13.8	6.2		

11- Barbital buffer stock solutions
- A: 0.2 M solution of sodium barbital (veronal) $C_8H_{11}N_2NaO_3$ (41.2 g in 1000 ml)
- B: 0.2 M HCl

50 ml of Ax ml of B, diluted to a total of 200 ml:

x	pH	x	pH
1.5	9.2	22.5	7.8
2.5	9.0	27.5	7.6
4.0	8.8	32.5	7.4
6.0	8.6	39.0	7.2
9.0	8.4	43.0	7.0
2.7	8.2	45.0	6.8
17.5	8.0		

Solutions more concentrated than 0.05 M may crystallize on standing, especially in the cold.

12- Boric acid–borax buffer stock solutions

- A: 0.2 M solution of boric acid BH_3O_3 (12.4 g in 1000 ml)
- B: 0.05 M solution of borax (sodium borate $Na_2B_4O_7.10H_2O$) (19.05 g in 1000 ml)
- 50 ml of A, x ml of B, diluted to a total of 200 ml:

x	pH	x	pH
2.0	7.6	22.5	8.7
3.1	7.8	30.0	8.8
4.9	8.0	42.5	8.9
7.3	8.2	59.0	9.0
11.5	8.4	83.0	9.1
17.5	8.6	115.0	9.2

13- 2-Amino-2-methyl-1,3-propanediol (Ammediol) buffer stock solutions

- A: 0.2 M solution of 2-amino-2-methyl-1,3-propanediol $C_4H_{11}NO_2$ (21.03 g in 1000 ml)
- B: 0.2 M HCl
- 50 ml of A, x ml of B, diluted to a total of 200 ml:

x	pH	x	pH
2.0	10.0	22.0	8.8
3.7	9.8	29.5	8.6
5.7	9.6	34.0	8.4
8.5	9.4	37.7	8.2
12.5	9.2	41.0	8.0
16.7	9.0	43.5	7.8

14- Glycine–NaOH buffer stock solutions
- A: 0.2 M solution of glycine $C_2H_5NO_2$ (15.01 g in 1000 ml)
- B: 0.2 M NaOH
- 50 ml of A, x ml of B, diluted to a total of 200 ml:

x	pH	x	pH
4.0	8.6	22.4	9.6
6.0	8.8	27.2	9.8
8.8	9.0	32.0	10.0
12.0	9.2	38.6	10.4
16.8	9.4	45.5	10.6

15- Borax–NaOH buffer stock solutions
- A: 0.05 M solution of borax (19.05 g in 1000 ml)
- B: 0.2 M NaOH
- 50 ml of Ax ml of B, diluted to a total of 200 ml:

X	pH
0.0	9.28
7.0	9.35
11.0	9.4
17.6	9.5
23.0	9.6
29.0	9.7
34.0	9.8
38.6	9.9
43.0	10.0
46.0	10.1

16- Carbonate–bicarbonate buffer stock solutions
- A: 0.2 M solution of anhydrous sodium carbonate (21.2 g in 1000 ml)
- B: 0.2 M solution of sodium bicarbonate (16.8 g in 1000 ml)
- x ml of A, y ml of B, diluted to a total of 200 ml:

X	y	pH
4.0	46.0	9.2
7.5	42.5	9.3
9.5	40.5	9.4
13.0	37.0	9.5
16.0	34.0	9.6
19.5	30.5	9.7
22.0	28.0	9.8
25.0	25.0	9.9
27.5	22.5	10.0
30.0	20.0	10.1
33.0	17.0	10.2
35.5	14.5	10.3
38.5	11.5	10.4
40.5	9.5	10.5
42.5	7.5	10.6
45.0	5.0	10.7

17- Glycine–HCl buffer

- A: 0.2 M solution of glycine (15.01 g in 1000 ml)
- B: 0.2 M HCl
- 50 ml of Ax ml of B, diluted to a total of 200 ml:

x	pH	x	pH
5.0	3.6	16.8	2.8
6.4	3.4	24.2	2.6
8.2	3.2		

18- Volatile buffers

In certain cases, it is necessary to remove a buffer quickly and completely. Volatile buffers make it possible to remove components that may interfere in subsequent procedures. It is useful in electrophoresis, ion-exchange chromatography, and digestion of proteins followed by separation of peptides or amino acids. Most of the volatile buffers are transparent in the lower UV range except for the buffers containing pyridine (Perrin & Dempsey, 1974). An important consideration is an interference in amino acid analysis (i.e., reactions with ninhydrin).

Most volatile buffers will not interfere with ninhydrin if the concentrations are not too high (e.g., triethanolamine less than 0.1 M does not interfere).

Effective pH range	Buffer
3.3 – 4.3	Formic acid
3.3 – 4.3	Pyridine / formic acid
3.3 – 4.3	Trimethylamine / formic acid
3.3 – 4.3	Ammonia / formic acid
4.3 – 5.3	Trimethylamine / acetic acid
4.3 – 5.3	Ammonia / acetic acid
4.3 – 5.3	N-ethylmorpholine / acetate
4.3 – 5.8	Pyridine / acetic acid
4.8 – 5.8	Pyridine / formic acid
5.9 – 6.9	Trimethylamine / carbonate
5.9 – 6.9	Ammonium bicarbonate
5.9 – 6.9	Ammonium carbonate / ammonia
5.9 – 6.9	Ammonium carbonate
6.8 – 8.8	Trimethylamine / hydrochloric acid
7.0 – 8.2	N-ethylmorpholine / acetate
8.8 – 9.8	Ammonia / formic acid
8.8 – 9.8	Ammonia / acetic acid
8.8 – 9.8	Ammonium bicarbonate
8.8 – 9.8	Ammonium carbonate / ammonia
8.8 – 9.8	Ammonium carbonate
9.3 – 10.3	Trimethylamine / formic acid
9.3 – 10.3	Trimethylamine / acetic acid
9.3 – 10.3	Trimethylamine / carbonate

Volatile buffer systems.

VII.   Formulas needed in buffer preparation

Molarity (M) =	$$\dfrac{no.\,of\;moles}{V(L)} = \dfrac{Wt}{M.Wt \times V(L)}$$ $$\dfrac{\%\times sp.gr \times 10}{M.Wt} = \dfrac{\%\times d \times 10}{M.Wt}$$	**Wt:** weight (gm) **M.Wt:** molecular weight **V: volume** **Sp.gr:** specific gravity **d:**density
Molality (m) =	$$\dfrac{no.\,of\;moles\;of\;solut}{Wt\,of\,solvent\,(Kgm)}$$ $$\dfrac{1000M}{1000d - M.Wt}$$	**M:** molarity **d:**density
Normality (N) =	$$\dfrac{no.\,of\,gram\,equivalent}{V(L)} = \dfrac{Wt \times n}{M.wt \times V(L)}$$ $$\dfrac{\%\times sp.gr \times 10 \times n}{M.Wt} = \dfrac{\%\times d \times 10 \times n}{M.Wt}$$	**n:** valency
% weight=	$$\dfrac{Wt\,of\,solute\,(gm)}{Wt\,of\,solvent\,(gm)} \times 100$$	
% volume =	$$\dfrac{volume\,of\,solute\,(ml)}{volume\,of\,solvent\,(gm)} \times 100$$	
Strength (S) =	$$\dfrac{Wt\,(gm)}{V(L)} = M \times M.Wt = N \times eq.Wt$$	
Parts Per Million (ppm) *concentration* **=**	$$\dfrac{Wt\,(m.gm)}{V(L)} = S \times 1000$$	

VIII. Molecular weight of some compounds

Compound	Structure	Molecular weight
2-amino-2-methyl-1,3-propanediol	$C_4H_{11}NO_2$	105.13
acetic acid	CH_3CO_2H	60.05
Anhydrous sodium carbonate	Na_2CO_3	105.98
Boric acid	BH_3O_3	61.83
Citric acid	C6H8O7 anhydrous	192.12
	$C6H8O7.(H_2O)$	210.13
Disodium hydrogen phosphate Or Dibasic sodium phosphate	Na_2HPO_4 anhydrous	141.95
	$Na_2HPO_4.2H_2O$	177.98
	$Na_2HPO_4 .7H_2O$	268.06
	$Na_2HPO_4. 12H_2O$	358.14
EDTA Ethylenediamine tetraacetic acid	$(HO_2CCH_2)_2NCH_2CH_2N(CH_2CO_2H)_2$	292.24
glycine	$C_2H_5NO_2$	75.06
Hydrocholic acid	HCL	36.46
Potassium chloride	KCl	74.551
Potassium dihydrogen phosphate	KH_2PO_4	136.08
Sodium acetate	$C_2H_3O_2Na$	82.03
Sodium barbital (veronal)	$C_8H_{11}N_2NaO_3$	206.17
Sodium	$NaHCO_3$	84.00

bicarbonate		
Sodium borate	$Na_2B_4O_7.10H_2O$	381.37
Sodium cacodylate	$(CH_3)_2AsO_2Na \cdot 3H_2O$	214.02
Sodium chloride	NaCl	58.44
Sodium dihydrogen phosphate Or monobasic sodium phosphate	NaH_2PO_4 anhydrous	119.97
	$NaH_2PO_4.H_2O$	137.99
	$NaH_2PO_4.2 H_2O$	156.00
Sodium hydroxide	NaOH	39.99
Tris (hydroxymethyl) aminomethane	$NH_2C(CH_2OH)_3$	121.14
Trisodium citrate	$C6H5O7Na_3. 2H_2O$	294.09

CHAPTER 2
SAMPLES COLLECTION IN SMALL LABORATORY ANIMALS

I. General principles of blood collection in animals

The method of blood collection must be described in the protocol approved by the Institute animal ethics committee.

It should be least painful and stressful so blood sample may be collected under anesthesia or without anesthesia.The training is required for blood collection using any method in any species.

If the study involves repeated blood sample collection, the samples can be withdrawn through a temporary cannula. This may reduce pain and stress in the experimental animals.

The acceptable quantity and frequency of blood sampling in all species is dependent upon the total blood volume of the animal. The estimated blood volume in adult animals is ≈55 to 70 ml/kg body weight. Care should be taken for older and obese animals.

Factors to consider in choosing the blood withdrawal technique include:

- The species to be bled
- The size of the animal to be bled
- The type of the sample required (e.g. serum, whole cells, etc.)
- The quality of the sample required (sterility, tissue fluid contamination, etc.)
- The quantity of blood required
- The frequency of sampling
- Health status of the animal being bled

- The training and experience of the technician

II. Replacement of Fluids

- Replace isotonic fluids (i.e. fluids with the same tonicity as blood) if >10% of total blood volume is required.

- if >10% blood volume is required, it is recommended to replace collected blood volume by 3–4 times the volume of blood collected with isotonic fluids (i.e. fluids with same tonicity as blood, such as 0.9% saline, 5% dextrose or Lactated Ringer's solution).

Species	Circulating blood volume (ml/kg bw)	20% (ml/kg BW)	15% (ml/kg BW)	10% (ml/kg BW)	7.5% (ml/kg BW)
Mouse	72	14.4	10.8	7.2	5.4
Rat	64	12.8	9.6	6.4	4.8
Rabbit	56	11.2	8.4	5.6	4.2
Non-human primate: Rhesus	56	11.2	8.4	5.6	4.2
Non-human primate: Cynomolgus	65	13.0	9.8	6.5	4.8
Non-human primate: Marmoset	71	14.2	10.6	7.1	5.3
Guinea pig	73	14.6	11.0	7.3	5.5
Hamster	78	15.6	11.7	7.8	5.8
Cat	56	11.2	8.4	5.6	4.2
Dog	85	17.0	12.8	8.5	6.4

Blood volume by species

Species	Site	volume	Repeat sampling (daily)	General anesthesia required
Mouse	Saphenous vein	Medium to large	Yes	No
	Tail vein or artery	Small	Yes	No
	Submandibular puncture	Medium to large	Yes	No
	Tail tip	1-2 drops	Yes	No
	Jugular vein	Large	Yes	Yes
Rat	Saphenous vein	Medium to large	Yes	No
	Tail vein or artery	Small to medium	Yes	No
	Jugular vein	Large	Yes	Yes
Gerbil and hamster	Lateral tarsal vein	Medium	Yes	No
	Jugular vein	Large	Yes	Yes
Guinea Pig	Saphenous vein	Medium	Yes	No
	Marginal ear vein	Small	Yes	No
	Jugular vein	Large	Yes	Recommended
Rabbit	Marginal ear vein or central ear artery	Large	Yes	Local anesthesia
	Jugular vein	Large	Yes	Recommended
	Femoral vein	Medium to large	Yes	No
	Cephalic vein	Medium to large	Yes	No

Common survival blood collection sites

III. General methods for blood collection

Blood samples are collected using the following techniques:

1. **Blood collection not requiring anesthesia:**

- Saphenous vein (rat, mice, guinea pig)
- Dorsal pedal vein (rat, mice)

2. **Blood collection requiring anesthesia (local/general anesthesia):**

- Tail vein (rat, mice)
- Tail snip (mice)
- Orbital sinus (rat, mice)
- Jugular vein (rat, mice)
- Temporary cannula (rat, mice)
- Blood vessel cannulation (rat, guinea pig, ferret)
- Tarsal vein (guinea pig)
- Marginal ear vein/artery (rabbit)

IV. Procedure for saphenous vein blood Sample collection

- The back of the hind leg is shaved until saphenous vein is visible.
- Hind leg is immobilized and slight pressure may be applied gently above the knee joint.
- The vein is punctured using a 20G needle and enough volume of blood is collected with a capillary tube or a syringe with a needle. The punctured site is compressed to stop the bleeding.
- Continuous sampling should be avoided and
- Collecting more than four samples in a day (24-hour period) is not advisable

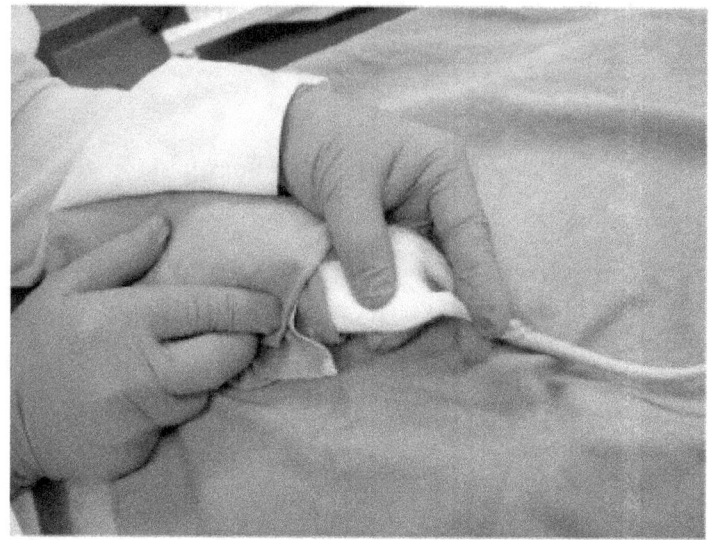

1- The animal is held in the restrainer head first so that only the rear legs and tail are free. The rear leg can be stretched out into a natural position.

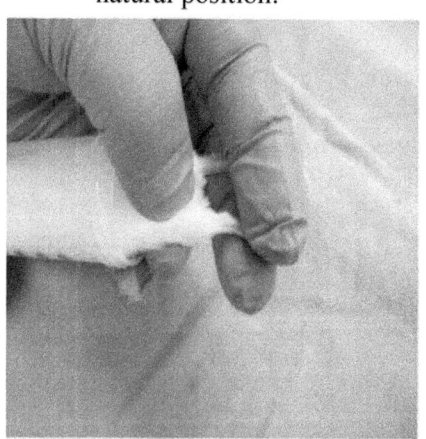 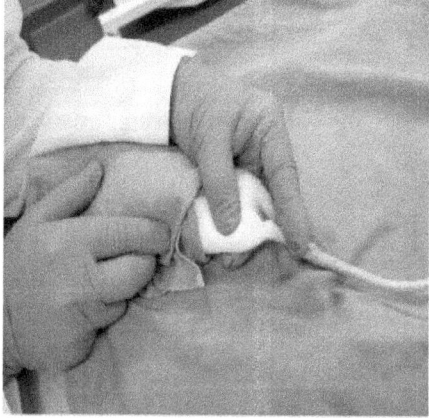

2- To secure the animal and elevate the vein, the skin on the upper thigh is gently but firmly squeezed, using the same hand that is holding the restrainer.

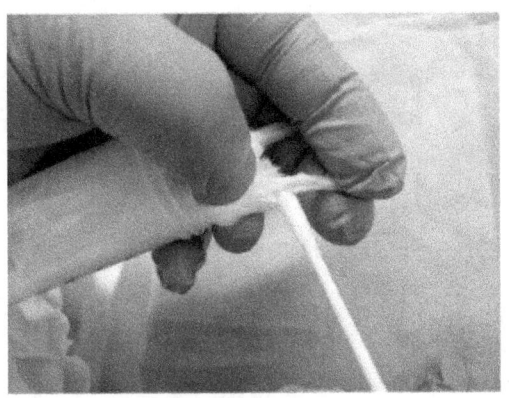 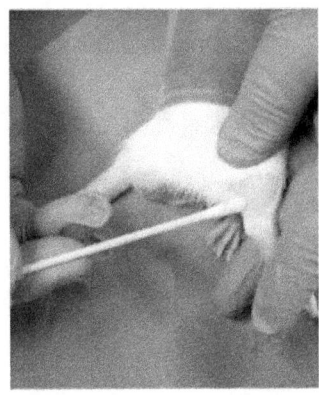

3- The hair is removed either by clippers or by using a depilatory cream and swabbing with 70% alcohol. Depilatory cream should not be left in contact with the skin for more than 1 minute and can be wiped away with alcohol. Swab the skin with a small amount of alcohol to help visualize the vein.

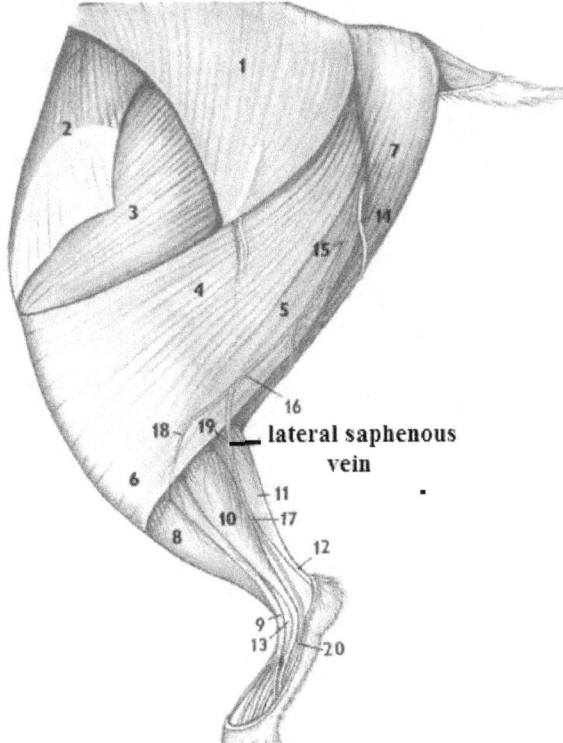

Position Locate the lateral saphenous vein

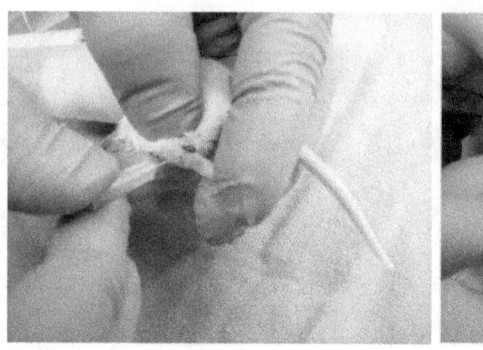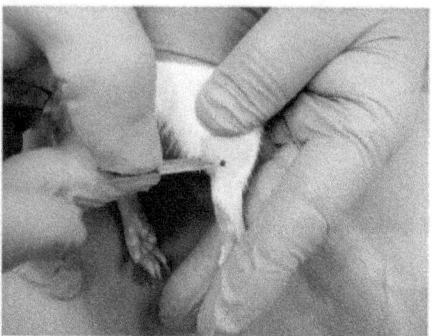

4- Using a 25-26 gauge needle or an animal lancet, puncture the vessel at a 90° angle at the most proximal (closest to the body) visible site.

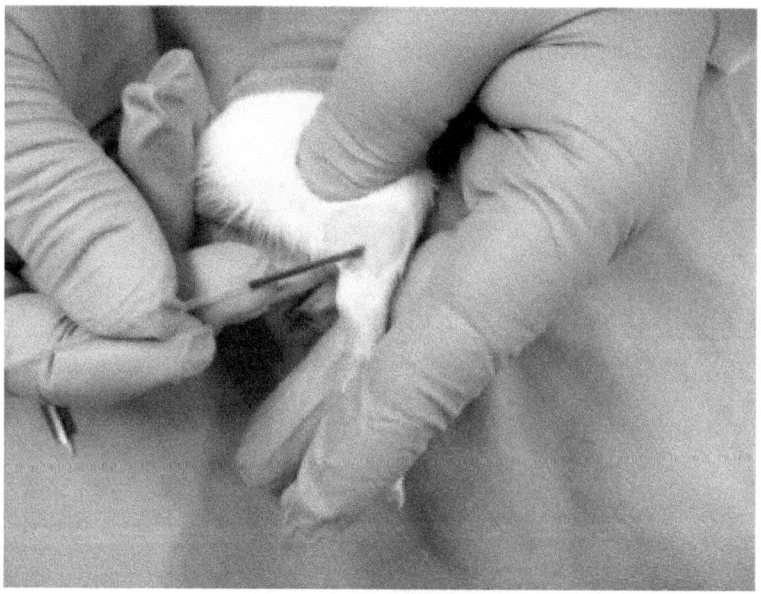

5- Collect the sample into your collection tube ensuring you do not exceed the allotted 10% blood volume loss.

V. Procedure for dorsal pedal vein blood sample collection

- The hind foot around ankle is held and medial dorsal pedal vessel is located on top of the foot.

- The foot is cleaned with absolute alcohol and dorsal pedal vein is punctured with 23G/27G needle.

- Drops of blood that would appear on the skin surface are collected in a capillary tube and a little pressure is applied to stop the bleeding .

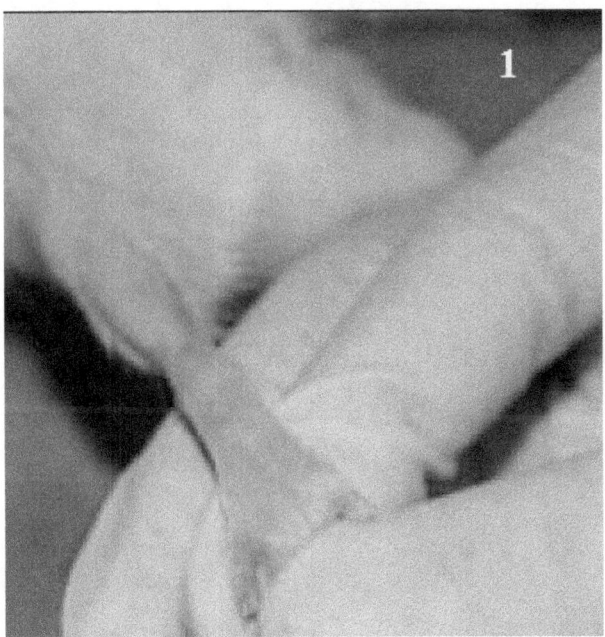

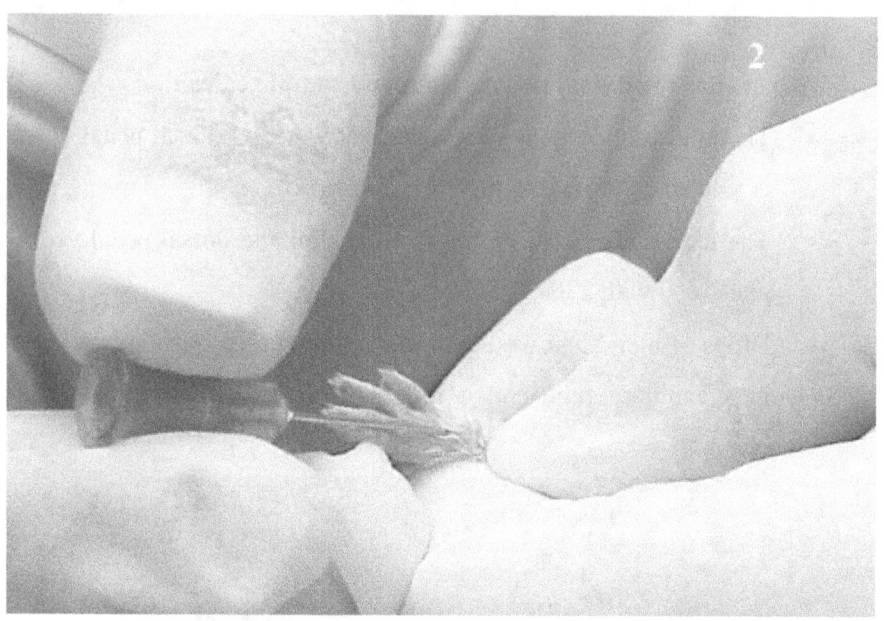

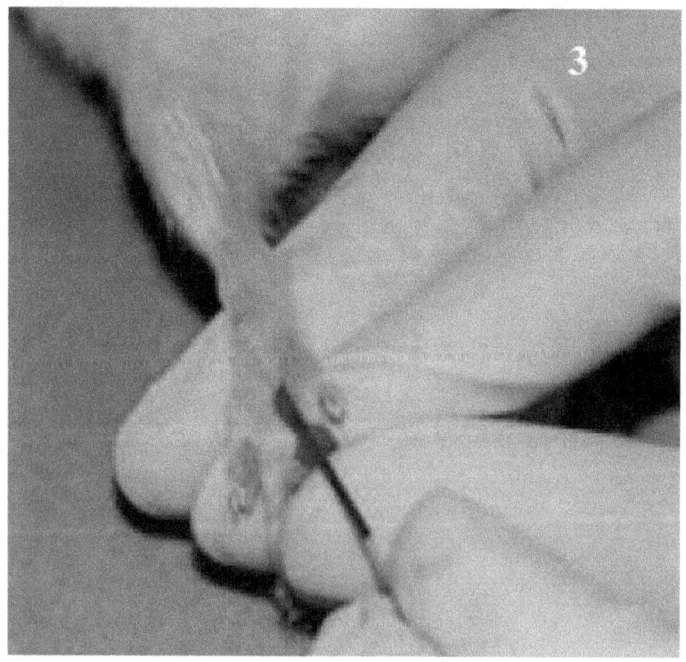

VI. Procedure for tail vein blood sample collection

- This method is recommended for collecting a large volume of blood sample (up to 2ml /withdrawal)

- The tail should not be rubbed from the base to the tip as it will result in leukocytosis. If the vein is not visible, the tail is dipped into warm water (40°C).

- Local aesthetic cream must be applied on the surface of the tail 30 min before the experiment.

- A 23G needle is inserted into the blood vessel and blood is collected using a syringe.

- In case of difficulties, 0.5 to 1 cm of surface of the skin is cut open and the vein is pricked with bleeding lancet or needle and blood is collected with a capillary tube or a syringe with a needle.

- Having completed blood collection, pressure/silver nitrate ointment/solution is applied to stop the bleeding.

- If multiple samples are needed, temporary surgical cannula may be used.

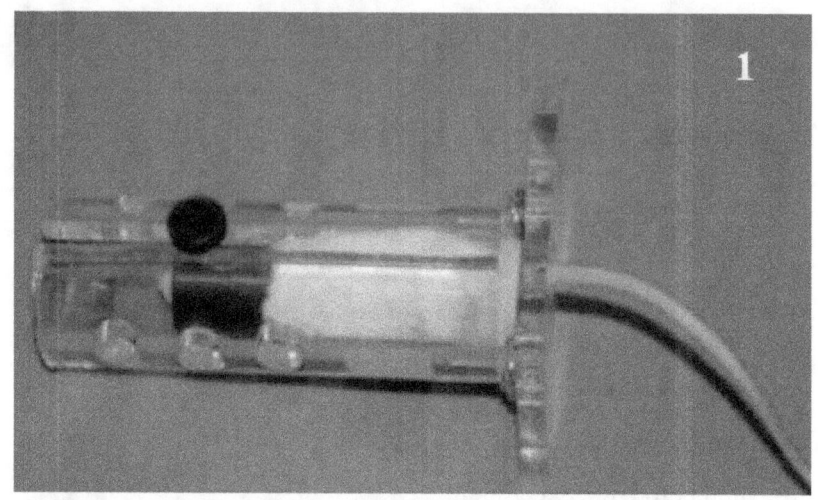

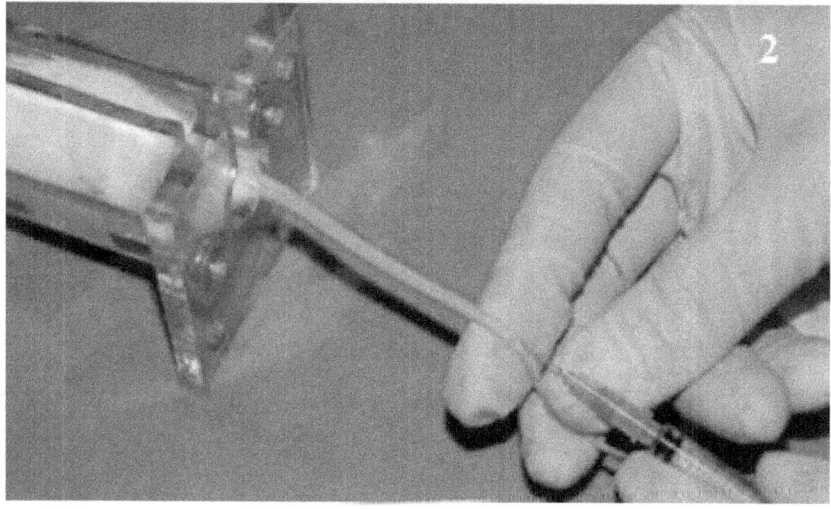

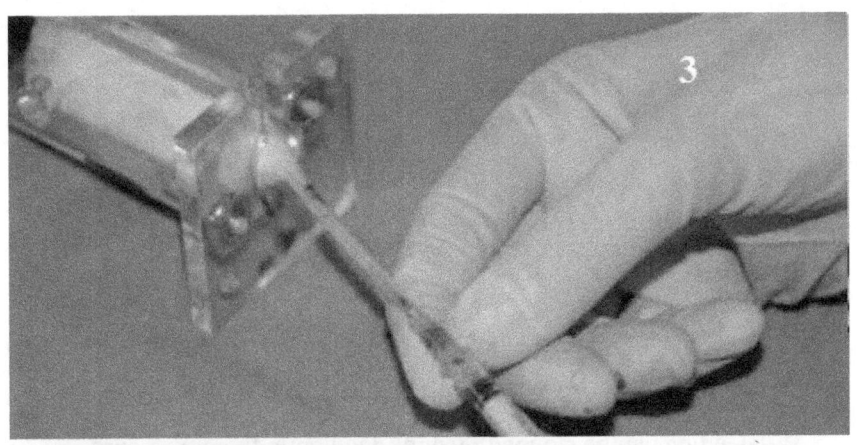

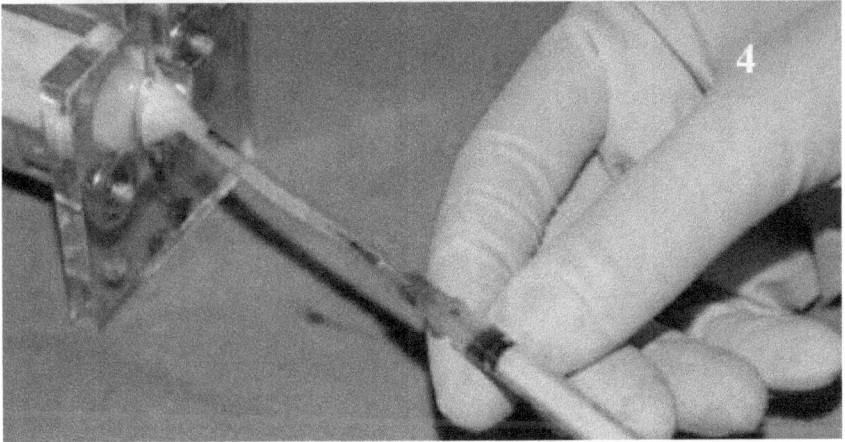

VII. Procedure for orbital sinus blood sample collection

- Requirements include animal, anesthetic agent, cotton, capillary tube and blood sample collection tubes.

- This technique is used with recovery in experimental circumstances and this method is also called periorbital, posterior-orbital and orbital venous plexus bleeding.

- Blood sample is collected under general anesthesia.

- Topical ophthalmic anesthetic agent is applied to the eye before bleeding.

- A capillary is inserted into the medial canthus of the eye (30 degree angle to the nose).
- Slight thumb pressure is enough to puncture the tissue and enter the plexus/sinus.
- Once the plexus/sinus is punctured, blood will come through the capillary tube.
- Once the required volume of blood is collected from plexus, the capillary tube is gently removed and wiped with sterile cotton. Bleeding can be stopped by applying gentle finger pressure.
- Thirty minutes after blood collection, animal is checked for postoperative and periorbital lesions

Caution:

- Repeated blood sampling is not recommended.
- Skill is required to collect blood.
- Even a minor mistake will cause damage to the eyes.
- Two weeks should be allowed between two bleedings.
- Adverse effects reported from this method is around 1 to 2% which includes hematoma, corneal ulceration, keratitis, pannus formation, rupture of the globe, damage of the optic nerve and other intraorbital structures and necrotic dacryoadenitis of the harderian gland.

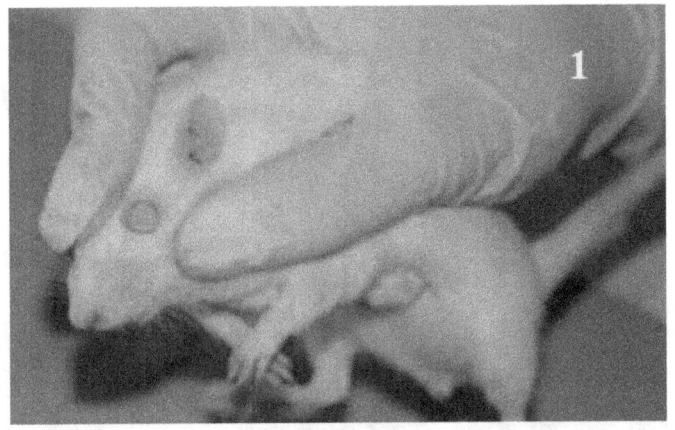

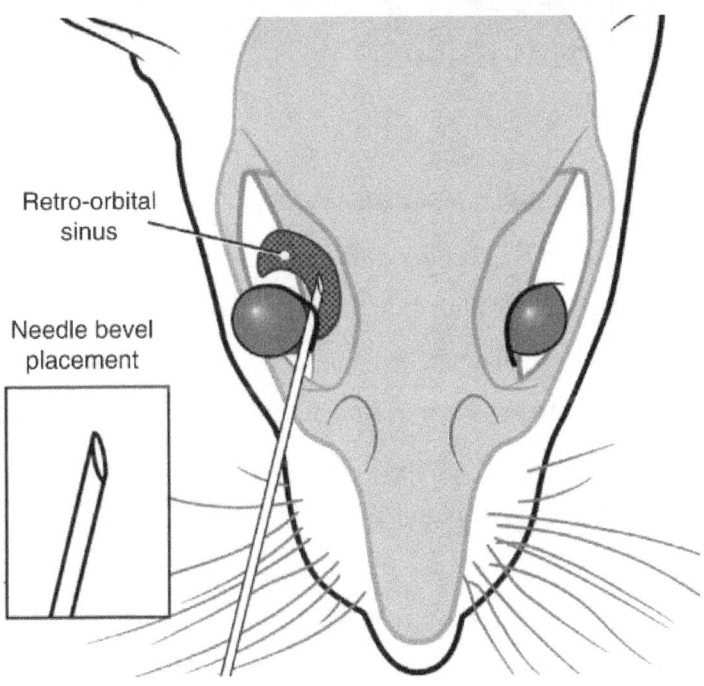

Retro-orbital
sinus

Needle bevel
placement

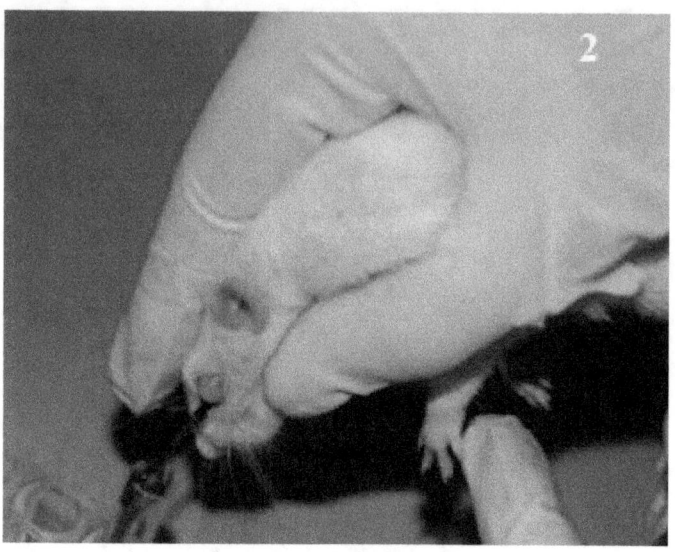

VIII. Procedure for jugular vein blood sample collection

- In this method, warming of the animals is not required and is used to collect micro volumes to one ml of blood sample.

- This method has to be carried out under general/inhalation anesthesia and two persons are needed to collect blood sample.

- One person has to restrain the animal and monitor the animal. Another person is required to collect the blood sample from the animal.

- The neck region of the animal is shaved and kept in hyperextended position. The jugular veins appear blue in color and is found 2 to 4 mm lateral to sternoclavicular junction. A 25G needle is inserted in the caudocephalic direction (back to front) and blood is withdrawn slowly to avoid collapse of these small blood vessels. Animal has to be handled carefully and not more than 3 to 4 mm of needle is to be inserted into the blood vessel.

- If the attempt to collect blood fails, the needle is slowly removed and the site is monitored for bleeding. If there is no bleeding, one more attempt can be made. Further attempts should be avoided in case of bleeding as it may collapse the vein.

- Finger pressure is applied to stop bleeding.

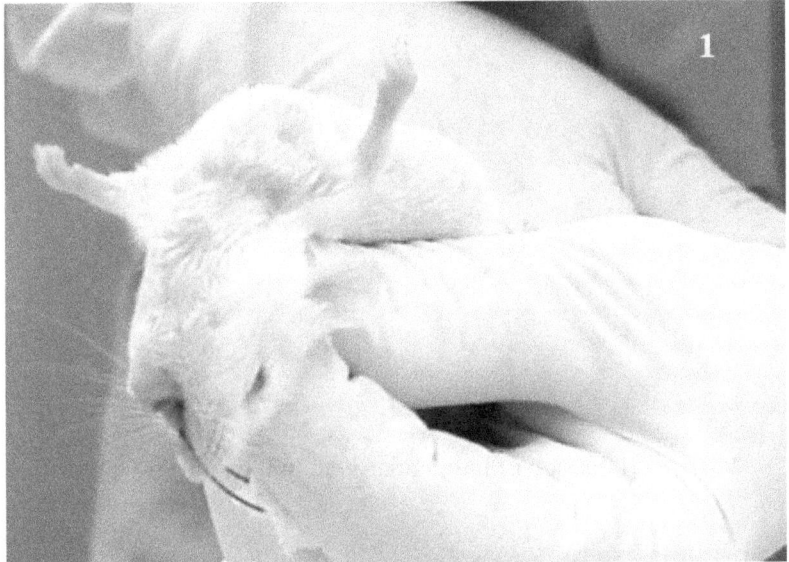

Proper positioning of mouse for blood collection from the jugular vein.

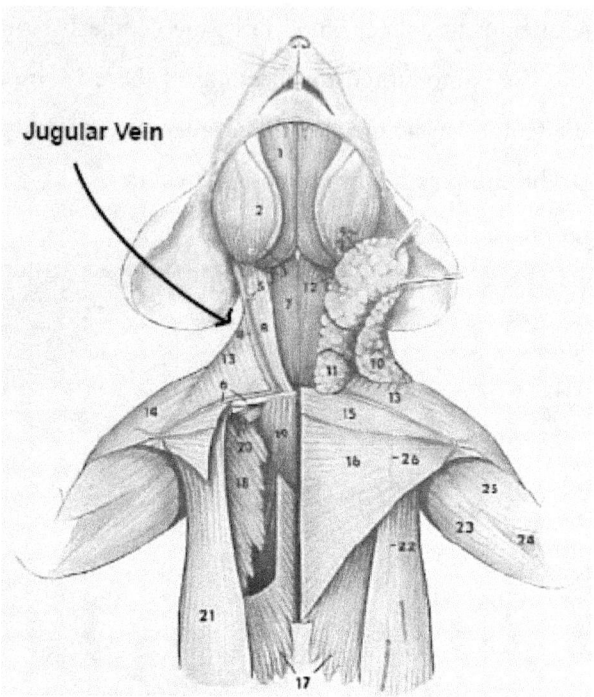

Jugular Vein

Location of the jugular vein in the mouse

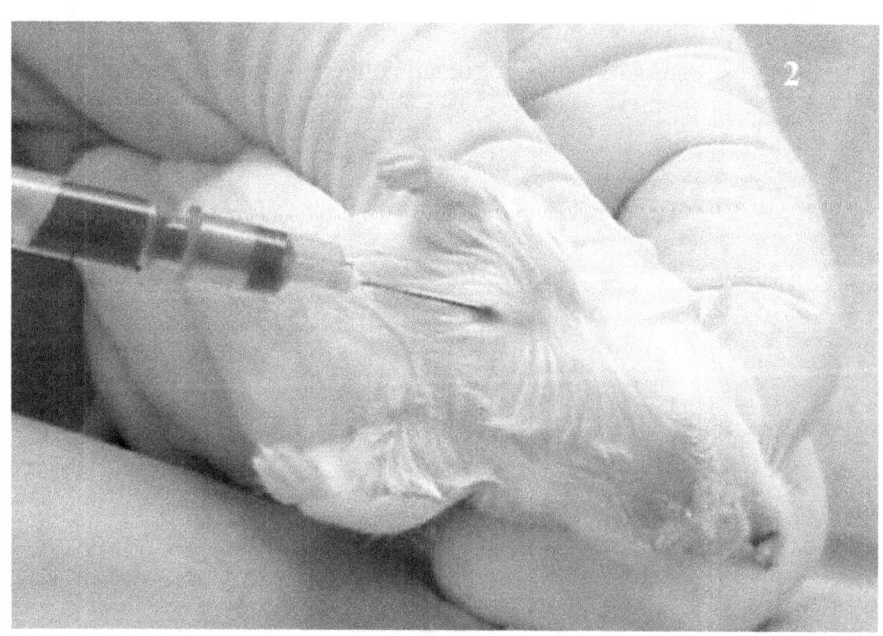

Blood collection from the jugular vein.

IX. Procedure for tarsal vein blood sample collection

- Tarsal vein is identified in one of the hind legs of large animals. This method is commonly recommended for guinea pig.
- One person has to restrain the animal properly.Tarsal vein may be visible in blue color.
- The surface hairs are removed by applying a suitable hair remover. A local anesthetic cream is applied on the collection site.
- After 20 to 30 minutes, blood sample is collected slowly by using 22G needle.
- Maximum three samples can be taken per leg and 0.1 to
- 0.3 ml of blood can be collected per sample. After the sample collection, gentle pressure is applied with finger for 2 minutes to stop bleeding.

Caution:

- Not more than six samples from both hind legs are taken.
- The number of attempts is three or less.

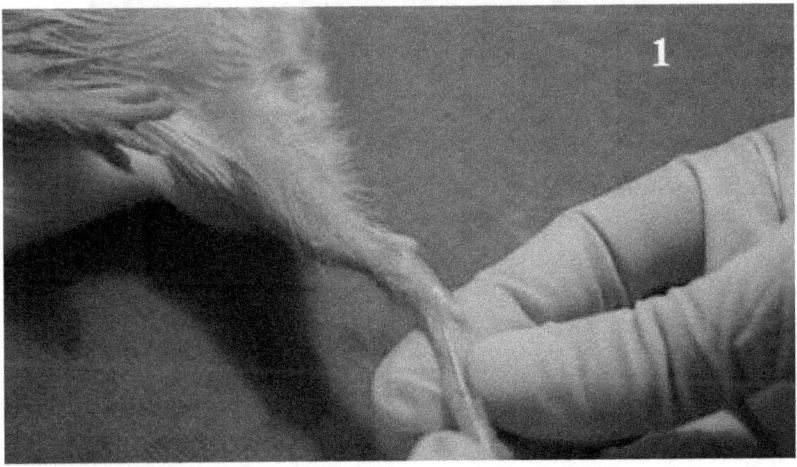

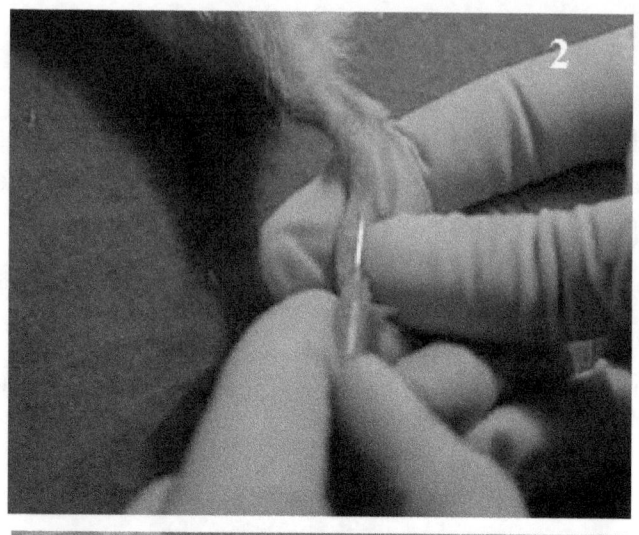

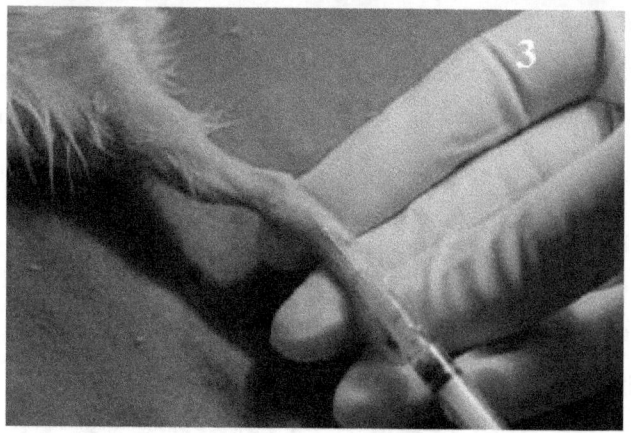

X. Termination Methods of Blood Collection Requiring Deep Anesthesia

1- Blood Collection by Cardiac Puncture

- Hold the mouse by the scruff of skin above the shoulders so that its head is up and its rear legs are down. Use a 1 ml syringe and a 22 gauge needle.

- Insert needle 5 mm from the center of the thorax towards the animal's chin, 5-10 mm deep, holding the syringe 25-30 degrees away from the chest
- Lay animal on back and push syringe vertically through sternum.
- If blood doesn't appear immediately, withdraw 0.5 cc of air to create a vacuum in the syringe. Withdraw the needle without removing it from under the skin and try a slightly different angle or direction.
- When blood appears in the syringe, hold it still and gently pull back on the plunger to obtain the maximum amount of blood available.
- Pulling back on the plunger too much will cause the heart to collapse.
- If blood stops flowing, rotate the needle or pull it out slightly

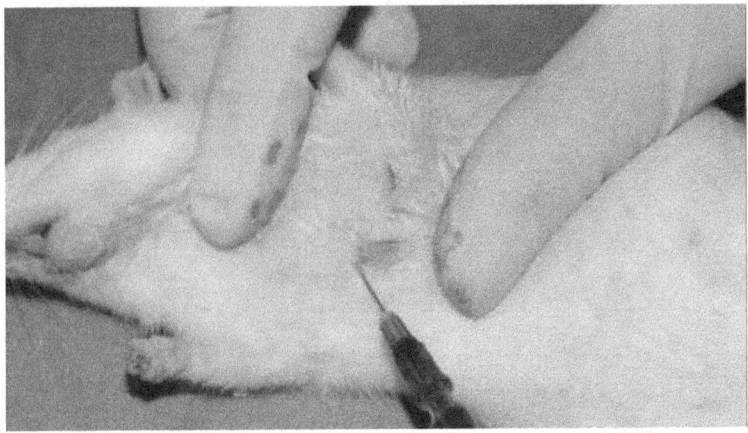

2- Blood Collection from the Posterior Vena Cava

- Open the abdominal cavity of anesthetized mouse by making a V-cut through the skin and abdominal wall 1 cm caudal to the rib cage.

- Shift the intestines over to the left and push the liver forward.
- Locate the widest part of the posterior vena cava (between the kidneys). Use a 23-25 gauge needle and a 1 ml syringe. Carefully insert the needle into the vein and draw blood slowly until the vessel wall collapses. Pause to allow the vein to refill and then repeat three or four times or until no more blood is available.

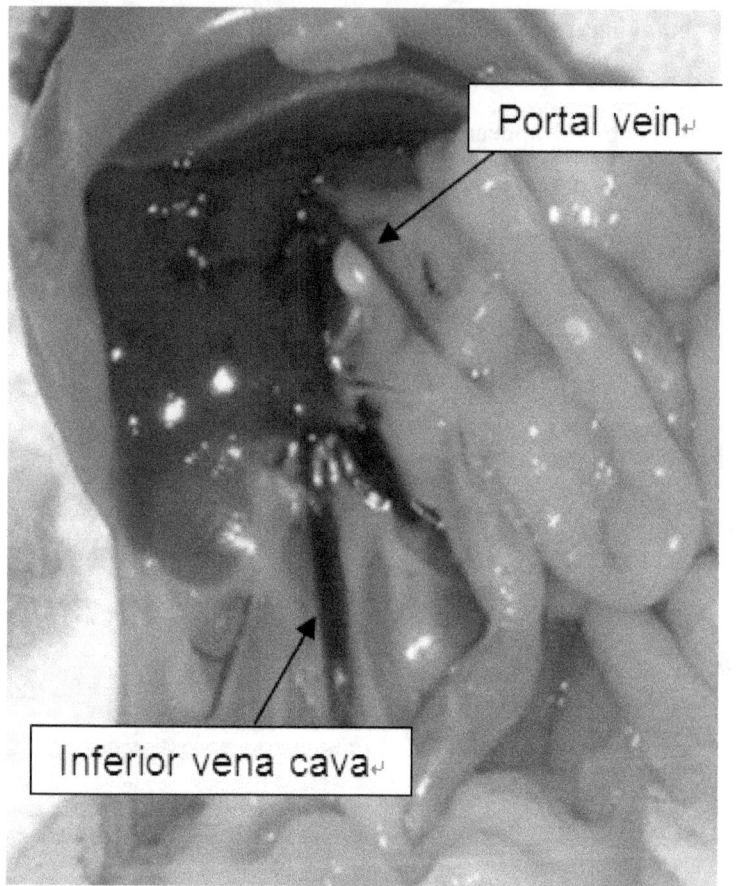

Portal vein and inferior vena cava *of mouse*

3- Blood Collection from the Axillary Vessels

- Lay the anesthetized mouse on its back.

- Stretch out a forelimb and pin the front foot.

- Make a deep incision in the axilla (armpit) at the side of the thorax.

- Hold the skin at the posterior part of the incision using forceps to create a cupped area. Incise the blood vessels in the area with a scalpel or straight edge razor and collect blood as it pools.

- It may be important to consider that tissue fluids will contaminate the blood sample.

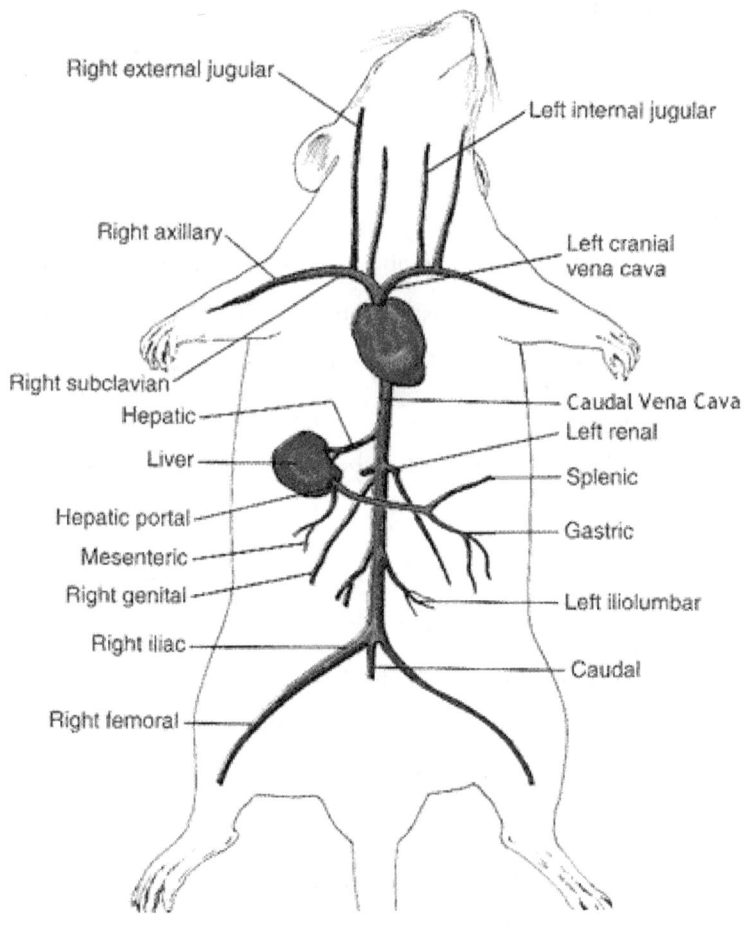

4- Blood Collection from the Orbital Sinus

- Quickly remove the eyeball from the socket with a pair of tissue forceps.

- Hold the mouse in the palm of your hand over a collection tube.

- Massage the body of the mouse with your hand by squeezing the rear half, then the middle, then the head of the mouse so that you are "milking" the blood toward the eye.

- When done properly, large drops of blood will flow from the orbital sinus.

XI. Euthanasia methods

The following methods are acceptable for use on mice and rats. Deviations from these procedures or the use of other methods of euthanasia require scientific justification and IACUC approval before they may be used.

Neonatal Rodents

- Neonatal rodents up to 10 days of age are resistant to euthanasia with CO_2.

- Euthanasia of neonatal rodents may be performed by CO_2 exposure followed by decapitation.

- Although neonates are resistant to volatile anesthetics and CO_2, prolonged exposure will induce anesthesia.

- Exposure to CO_2 or volatile anesthetics such as halothane and isoflurane should not be relied on as the sole method of euthanasia for neonates. It should be followed by decapitation.

1- CO_2 method

• Acceptable under specific conditions.

• The animal is placed into a closed chamber in which carbon dioxide gas is rapidly introduced.

• Compressed CO_2 gas in cylinders is the only acceptable source of carbon dioxide because the inflow to the chamber can be regulated precisely.

• Sexes should be separated and chambers should not be overcrowded.

• Do not combine unfamiliar male mice in a cage prior to euthanasia.

• Gas flow should be maintained for at least one minute after respiration ceases.

• It is important to verify that an animal is dead (i.e., no spontaneous respiration, pale or blue mucous membranes) before removing it from the chamber.

• Adjunctive methods of euthanasia should be used to assure death after exposure to CO_2:

   a) Cervical dislocation

   b) Decapitation (neonates)

   c) Stab incision between the ribs (each side of animal) to puncture the chest cavity and insure that the animal cannot respire.

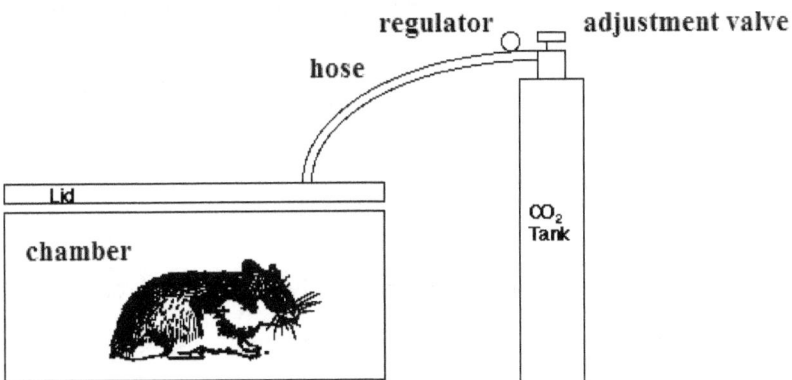

2- Cervical Dislocation

- Acceptable with the following conditions:
- Cervical dislocation may be performed on unconscious rodents.
- Investigators are responsible to determine that personnel using cervical dislocation are properly trained to do so.
- Cervical dislocation on conscious rodents requires scientific justification and prior approval by the IACUC.

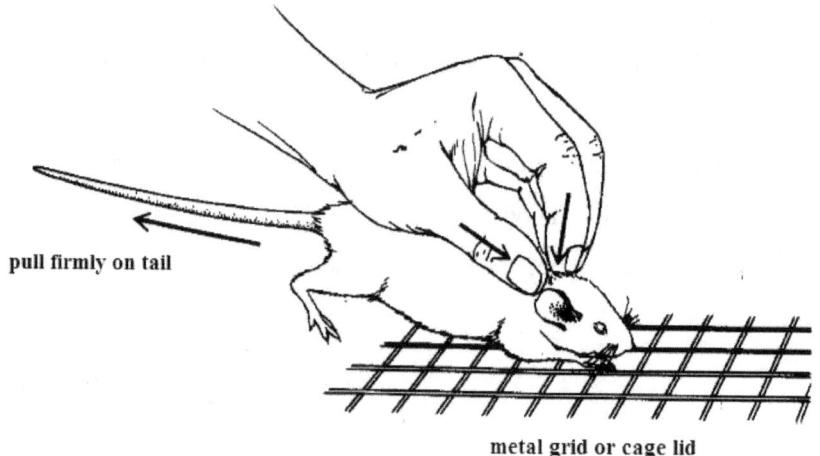

pull firmly on tail

metal grid or cage lid

3- Injectable Anesthetics

• Acceptable

• Anesthetics may be injected intraperitoneally to euthanize rodents.

• Agents available for use include sodium pentobarbital, other pentobarbital combinations and ketamine/xylazine combinations.

4- Inhalant Anesthetics

- Acceptable under specific conditions for euthanasia of laboratory rodents.
- A number of inhalant anesthetics may be used for anesthesia.
- All inhalant anesthetics require some method of scavenging the waste anesthetic vapors (i.e., working in a biosafety cabinet).

5- **Exsanguination**

- Acceptable only under deep anesthesia.
- May only be used to euthanize unconscious animals.

XII. Faces and urine collection (Metabolic cages)

- Metabolic cages allow separate collection of urine and faces from rodents for analysis of administered compounds and/or their metabolites.
- They are also useful for studies aiming at determining the excretion route utilized in the drug elimination.
- The rats were placed in metabolic cages in order to collect 24-h urine volume one day prior to sacrifice.
- The urine specimens were centrifuged and placed in a freezer at -20°C and the

Notes in using metabolic cage:

1- Rats should be able to be in visual, auditory and olfactory contact with other rats as far as possible.

2- Rats should be acclimatized to the metabolism cage before studies commence.

3- Where metabolism cages have to be used, consideration should be given to enriching the cages (for example with an area of solid floor and a nest box), providing this does not interfere with the study

XIII. Blood collection

1- Anticoagulants

- Additives that inhibit blood and/or plasma from clotting ensuring that the constituent to be measured is non-significantly changed prior to the analytical process.

- Anticoagulation occurs by binding calcium ions (EDTA, citrate) or by inhibiting thrombin activity (heparinates, hirudin).

- The following solid or liquid anticoagulants are mixed with blood immediately after sample collection:

EDTA (Ethylene diamine tetra acetic acid)

- Mode of action
  Form insoluble Ca salts
- Amount required
  10-20mg (1ml of 1 % solution)
- Advantage
  Recommended for routine hematological procedures, preserve cellular elements better
- Disadvantage
  May shrinks cell because Na salt is less soluble

Heparin

- Mode of action
  Antithrombin and antithromboplastin
- Amount required

  1-2mg (0.2ml of 1% solution)

- Advantage

  Less effect on RBC hemolysis

  Used for blood gas analysis

- Disadvantage

  May cause clumping of WBC,

  Unsuitable for smears, as it interferes with stain ability of WBC

  expensive

Na citrate

- Mode of action
  Form insoluble Ca salts
- Amount required

  10-20mg (1ml)

- Advantage

  Can be used for blood transfusion

- Disadvantage

  Interferes with many chemical tests, shrink cells

Potassium oxalate

- Mode of action
  Form insoluble Ca salts

- Amount required

  20mg (2 drops of 20% solution)
- Advantage

  Very soluble
- Disadvantage

  Causes shrinkage, it increase the volume of blood

Sodium oxalate

- Mode of action
  Form insoluble Ca salts
- Amount required

  20mg (2 drops of 20% solution)
- Advantage

  Used mainly for prothrombin time
- Disadvantage

  May shrinks cell because Na salt is less soluble

2- Vacutainer tubes

1- Red-stopper tubes, are for tests requiring clotted blood

2- Lavender stopper tubes, contain EDTA in concentrated liquid or desiccated powder form

3- Green stopper tubes, contain heparin and are used for blood gases, PH, (CO_2, O_2)....

4- Gray stopper tubes, contain oxalates, fluorides, or citrates

5- Yellow stopper tubes, available with Acid Citrate Dextrose (ACD) solution or physiological saline solution

3- Serum sampling

- Blood samples were collected directly in centrifuge tubes without anticoagulants and kept for 30 min at room temperature to clot.
- The clotted blood was then centrifuged at 1500 rpm for 10 min at room temperature.
- Serum was separated and then quickly stored at -80 °C.

4- Plasma sampling

- Centrifuge the anticoagulated blood (citrated, EDTA or heparinized blood) for at least 15 minutes at 2000 to 3000 g to obtain cell-free plasma.

5- Erythrocyte sampling

- The anticoagulated blood samples were centrifuged at 1000 x g for 10 min at 4 °C and the upper phase was taken with a Pasteur pipette into an eppendorf tube and stored at –40 °C (for plasma analysis).
- The buffy coat on top of the erythrocyte layer was carefully removed and 10 mL isotonic NaCl solution was added.
- Resuspended erythrocyte was centrifuged at 1000 x g for 10 min and the upper part removed again.
- Then 10 mL phosphate buffer solution (PBS) was added and the erythrocytes were centrifuged, and the upper buffer part removed by pasteur pipette.

- The erythrocytes were diluted 10 times with ice cold water, vortexed and stored at –40 °C until used.

XIV. Preparation of tissue Homogenate for oxidative stress biomarkers

- The liver was cleaned of blood using clean tissue paper

- Tissues were rinsed in ice-cold NaCl (0.9% w/v)

- Liver tissue was homogenized (10%, w/v) in ice-cold buffer:

    a) Tris-HCl (0.1M, pH 7.4) or

    b) 100 mM KH_2PO_4 buffer containing 1 mM EDTA (pH 7.4)

    or

    c) Phosphate Buffer Solution (50mM, pH7.4)
    d) 1.15% KCl and 0.05 M phosphate buffer pH 7.4

- Note: these only some common examples for homogenization buffers

- The tissue homogenates were centrifuged at 15,000 g for 15 min at 4°C.

- The supernatant was stored at -70°C

CHAPTER 3
HEMATOLOGY TECHNIQUES

I. **The neubabuer chamber, or hemocytometer**

The Neubauer chamber is a thick crystal slide with the size of a glass slide. (30 x 70 mm and 4 mm thickness). In a simple counting chamber, the central area is where cell counts are performed. The chamber has three parts. hemacytometer counting chamber to count blood cells (to count WBC, RBC, and Platelets, as well as, counting cells in other body fluids, e.g. CSF and semen analysis).

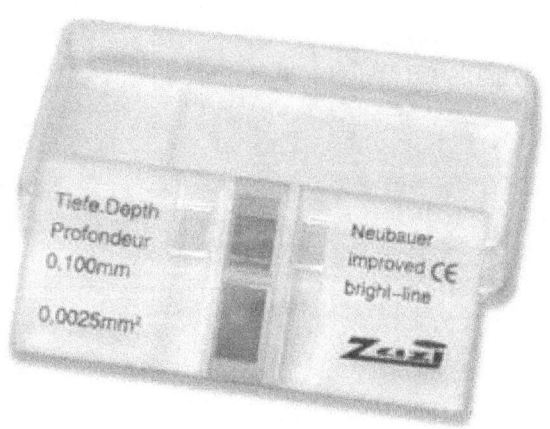

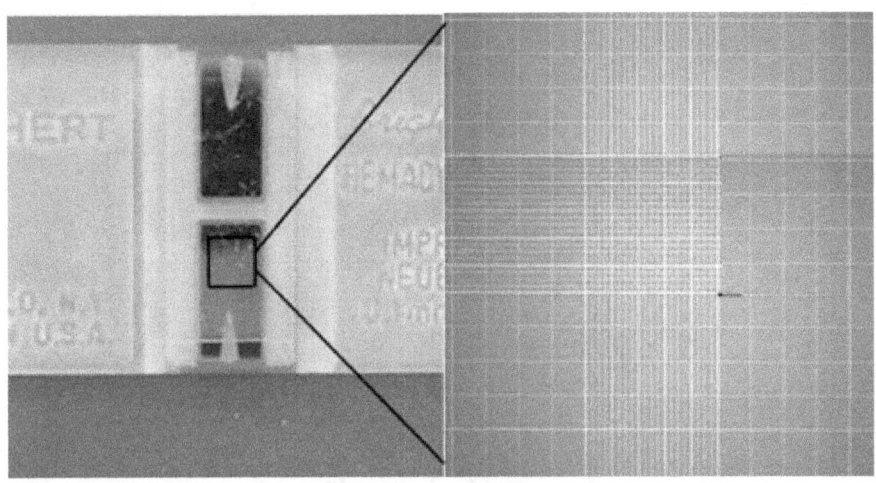

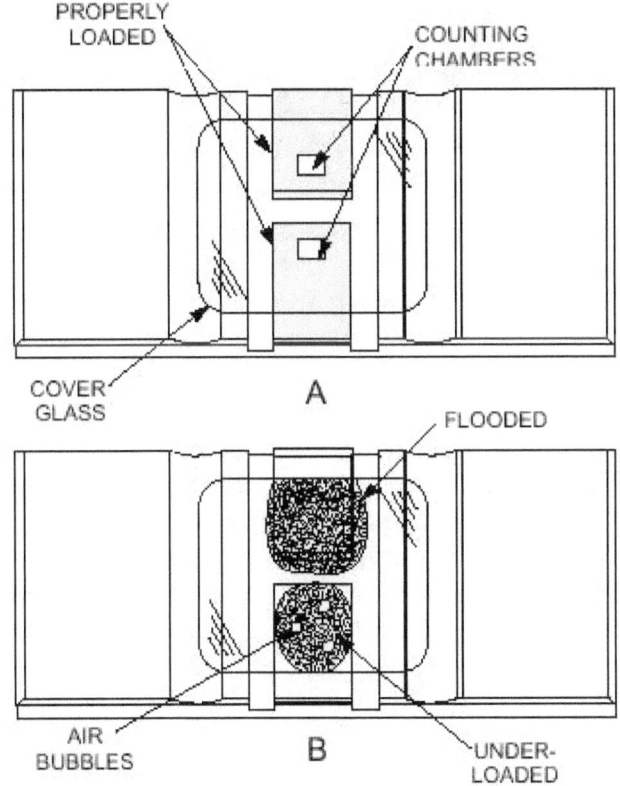

Neubauer chamber's counting grid is 3 mm x 3 mm in size. The grid has 9 square subdivisions of width 1mm.

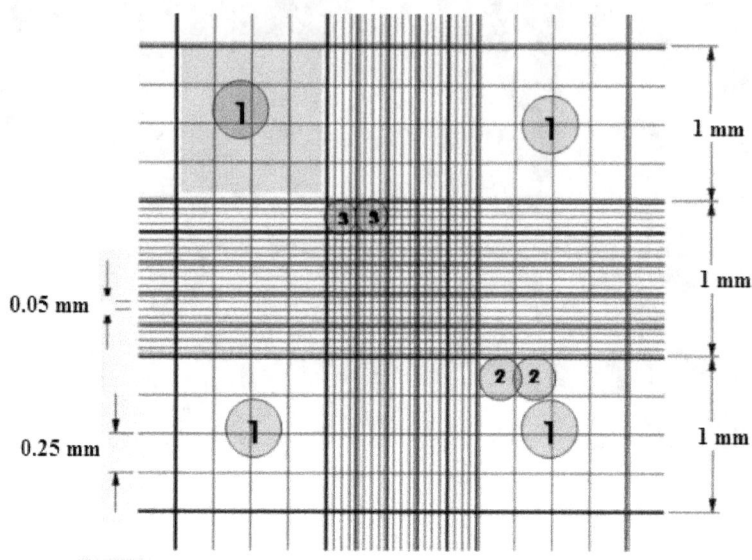

0.05 mm

0.25 mm

1 mm

1 mm

1 mm

1.   Large square (1mm x 1mm)

2.   Medium square (4 x 4 grid)

3.   Medium square central (5 x 5 grid)

In case of blood cell counting, the squares placed at the corners are used for white cell counting. Since their concentration is lower than red blood cells a larger area is required to perform the cell count. The central square is used for platelets and red cells. This square is split in 25 squares of width 0,2 mm (200μm). Each one of the 25 central squares is subdivided in 16 small squares. Therefore, the central square is made of 400 small squares.

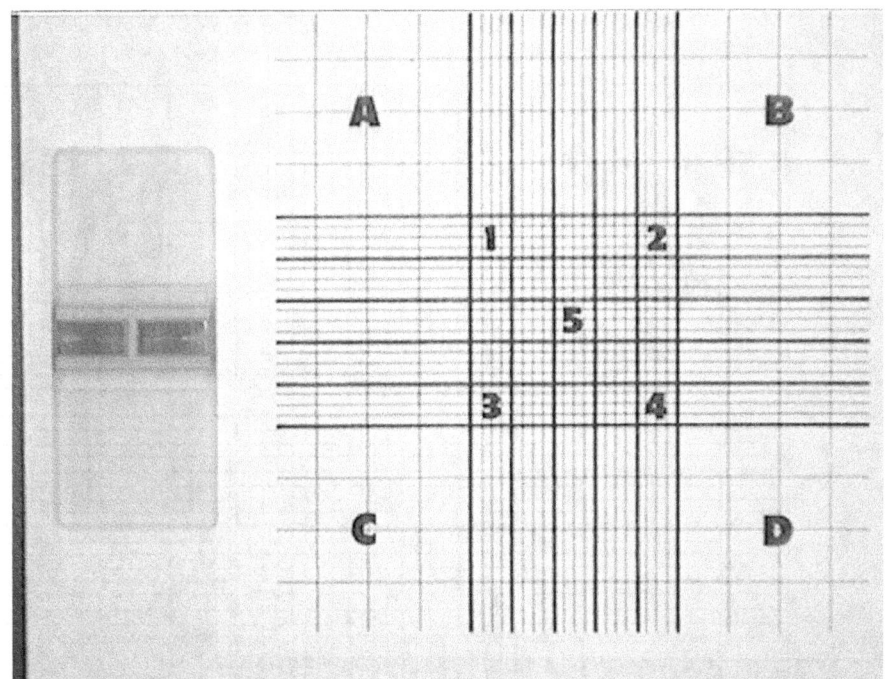

The cross-hatched central square is used for RBC counts: only the 4 small corner squares and the small center square are used for RBC counting (numbered squares). The large peripheral corner squares (areas containing letters A-D) are used for WBC counts.

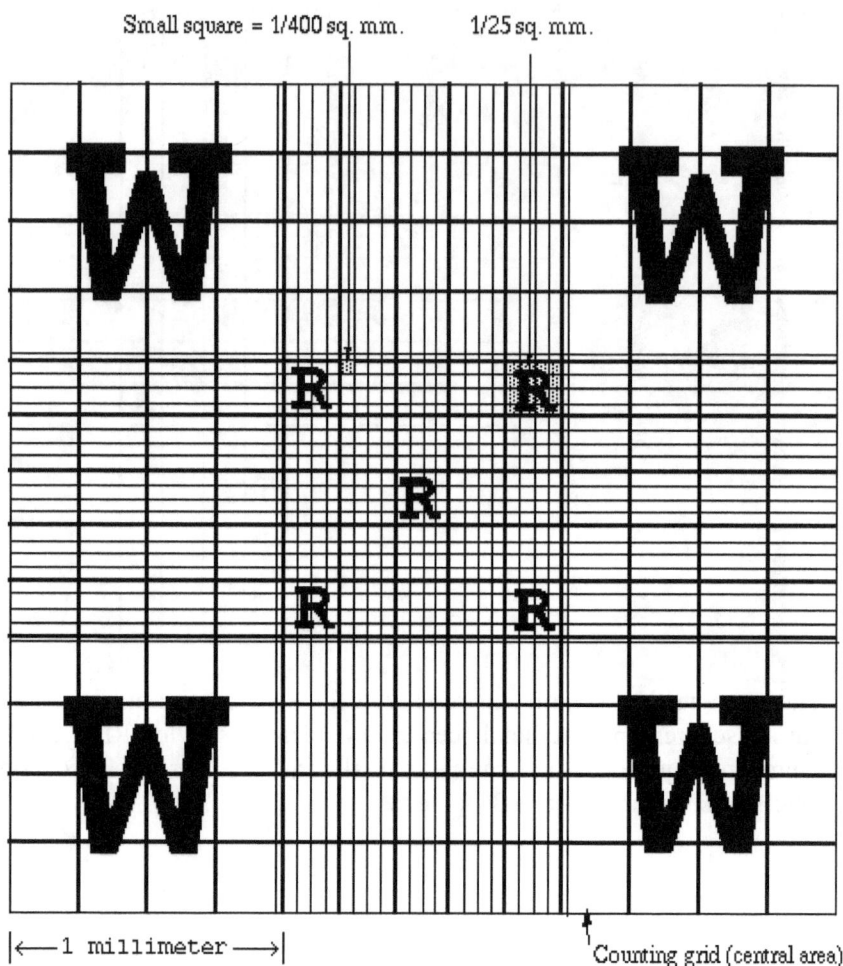

Small square = 1/400 sq. mm.   1/25 sq. mm.

|←—1 millimeter —→| Counting grid (central area)

II. Determination of total erythrocyte count by Haemocytometry

The total erythrocyte count was determined by haemocytometry following the method of Gottfried and Gerard.

Procedure:

1. Take the blood in to RBC pipette up to 0.5 marks
2. Immediately draw the RBC diluting fluid (normal Saline) up to mark 101.
3. Rotate the pipette between thumb and other fingers with finger eight (8) movements. This gives a dilution of **1:200**.

Nowadays, glass pipettes have been replaced by micropipettes.

4. Clean the counting chamber of hemocytometer and cover slip

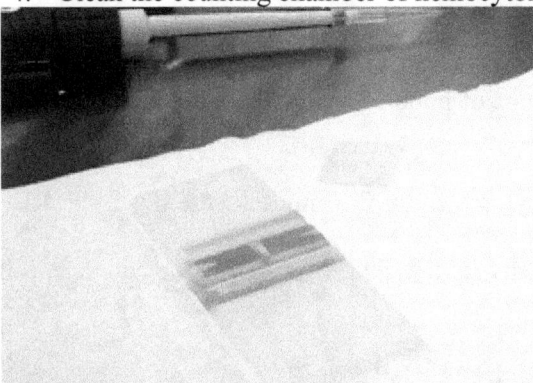

5. Put the glass cover on the Neubauer chamber central area.

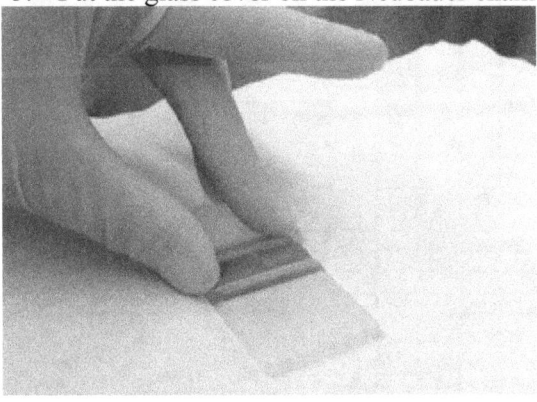

6. Suck 10 µl by micropipette
7. Place pipette tip close to the glass cover edge, right at the centre of the Neubauer chamber.
8. Release the plunger slowly watching how the liquid enters the chamber uniformly, being absorbed by capillarity.

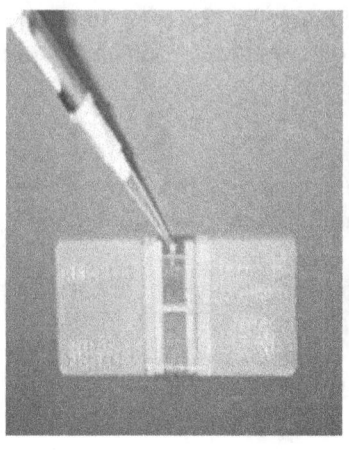

In case of the appearance of bubbles or that the glass cover has moved, repeat the operation.

9. Allow the hemocytometer for 2-3 min to settle down the RBC in counting chamber
10. The red blood cells in the four corner squares and one central square were counted.

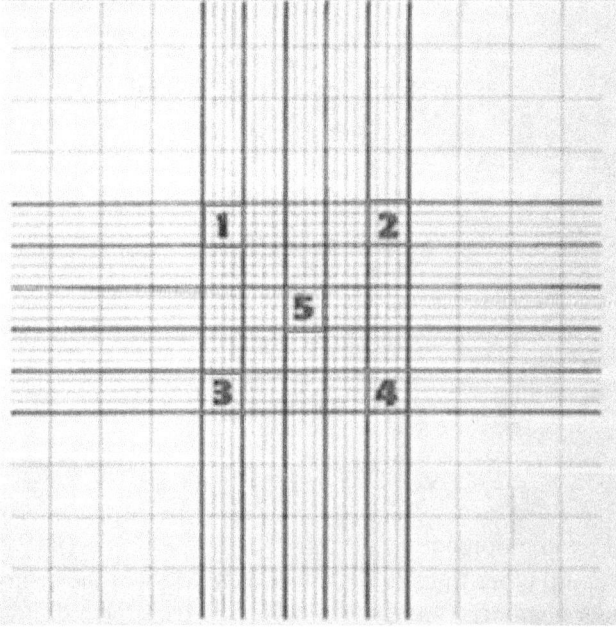

Counting rules:
- Count cells touching the **left and top** side lines.
- Don't count cells touching the **bottom right** side lines.
- Count first left to right direction, then to vise verse.

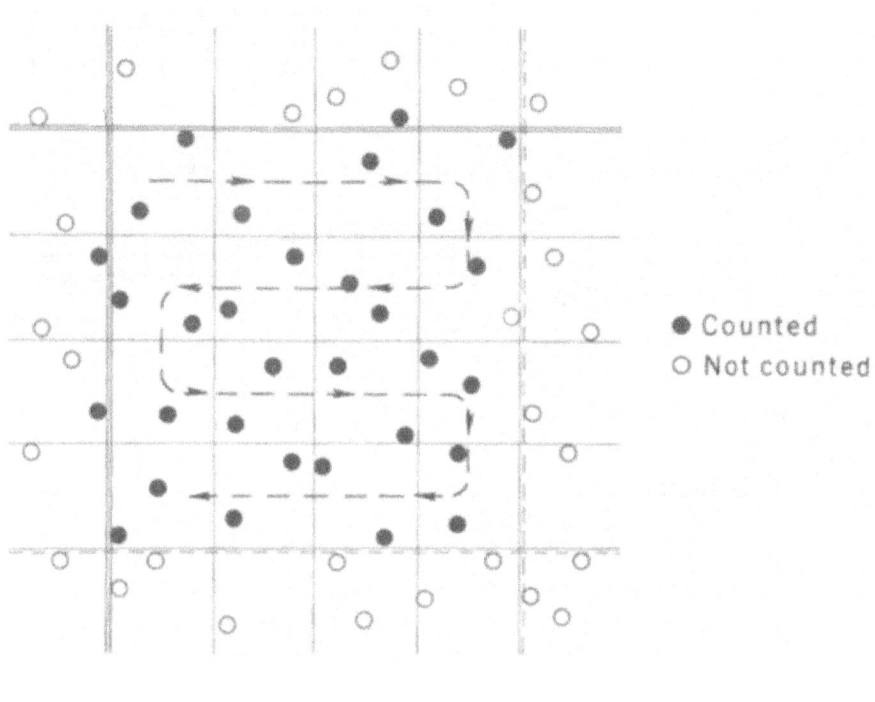

● Counted
○ Not counted

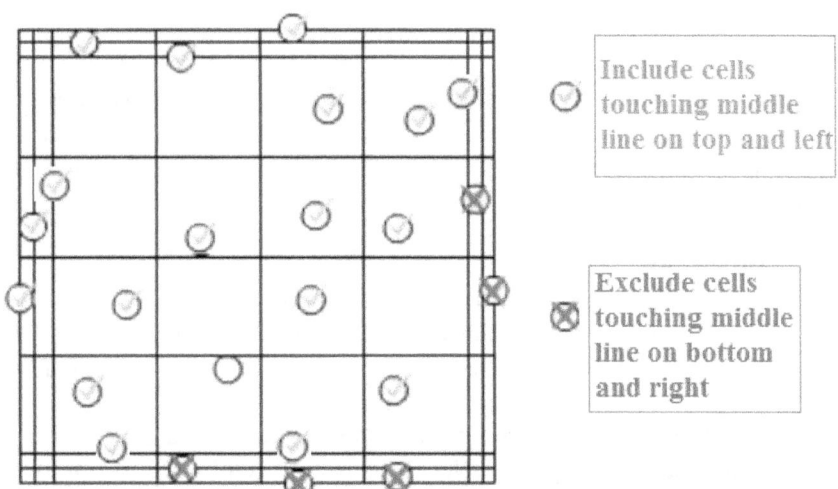

Include cells touching middle line on top and left

Exclude cells touching middle line on bottom and right

Calculation:

Volume of one small square $= \frac{1}{20}$ mm $\times \frac{1}{20}$ mm $\times \frac{1}{10}$ mm $= \frac{1}{4000}$ mm^3

Volume of 80 (5x16) small square $= 80 \times \frac{1}{4000}$ mm$^3 = \frac{1}{50}$ mm^3

Total number of RBC $= \dfrac{\text{Cells counted (N)}}{\text{Volume of all squares} \times \text{dilution factor}}$

Total RBC $= \dfrac{(N)}{\frac{1}{50} \times \frac{1}{200}}$ $= N \times 10{,}000$

RBC diluting fluid
1- Normal saline (0.9% NaCl)
2- Haymen's diluting fluid:
- Sodium chloride – 0.5% (to maintain osmolarity)
- Sodium Sulphate - 2.5% (to prevent aggregation of RBC)
- Mercuric Chloride - 0.25% (preservative)

III. Determination of total leucocyte count by Haemocytometry

The total leucocyte count was determined by haemocytometry following the method of Gottfried and Gerard.

Procedure:
1. Draw the blood in to WBC pipette up to 0.5 marks.
2. Immediately draw the WBC diluting fluid up to 11 marks.
3. Rotate the pipette between thumbs and finger horizontally. This will gives you a dilution of **1:20**.
4. Let stand for 10 minutes to allow red cells to hemolyzed.
Nowadays, glass pipettes have been replaced by micropipettes.
5. Clean the counting chamber of hematometer and cover slip.

6. Place the cover slip on the counting chamber with gentle pressure.

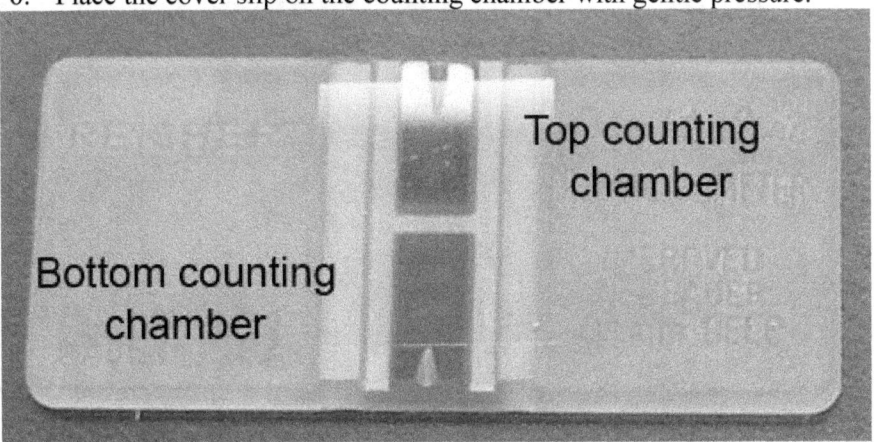

Top counting chamber

Bottom counting chamber

• Allow the hemocytometer for 2 minute to settle down the WBC.

• Count the WBC in the 4 large squares in the corners of counting chamber (16 small squares).

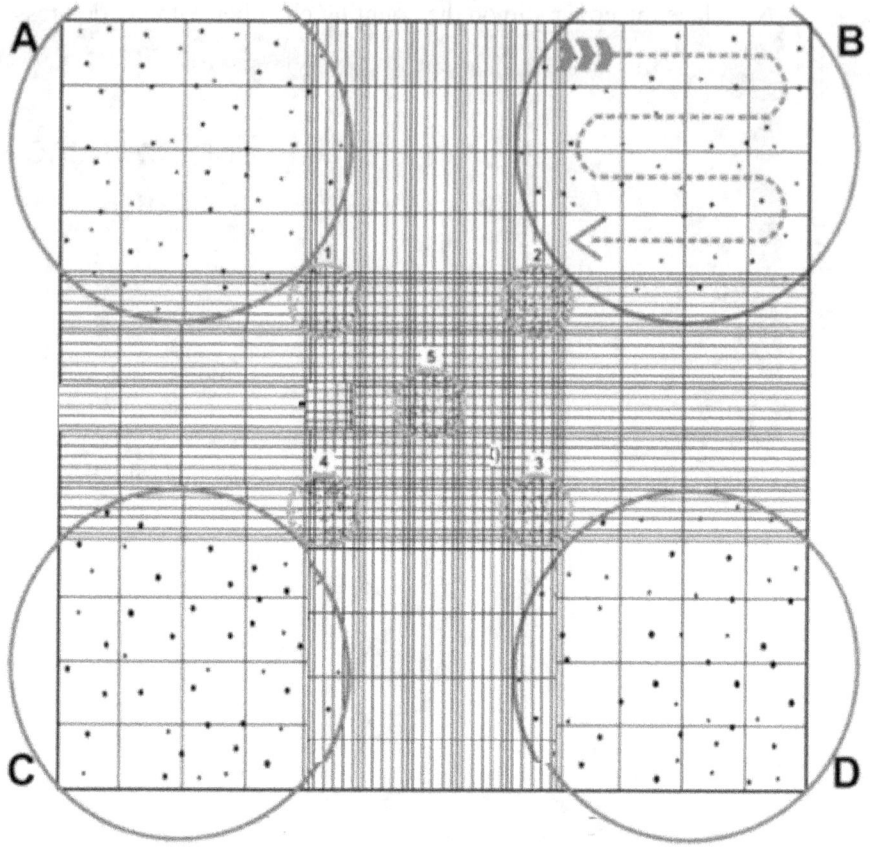

Counting rules:
Count cells touching the left and top side lines
Don't count cells touching the right and bottom side lines

Calculation:

Volume of one big squares = $\frac{1}{4}$ mm × $\frac{1}{4}$ mm × $\frac{1}{10}$ mm = $\frac{1}{160}$ mm³

Volume of 64 (4x16) big squares = 64 × $\frac{1}{160}$ mm³ = $\frac{4}{10}$ mm³

$$\text{Total number of WBC} = \frac{\text{Cells counted (N)}}{\text{Volume of all squares} \times \text{dilution factor}}$$

$$\text{Total WBC} = \frac{(N)}{\frac{4}{10} \times \frac{1}{20}} = N \times 50$$

Sample dilution:

Diluent for white blood cells:

- 10 mg crystal violet (gentian violet)
- 1.0 ml glacial acetic acid
- 100 mL d H_2O

Acetic acid facilitates haemolysis of RBC and Gentian violet stains the nuclei of WBC.

IV. Differential Leukocyte Count

1- Clean glass slide, sterile needle, microscope, stains solution, oil immersion
2- Making the smear
   - Place a small drop of blood near an end of the slide
   - Bring another slide in contact with the drop and allow distributing at an angle of 30-40 degrees
   - Push to the left in a smooth and quick motion
   - Dry the slide with air or methyl alcohol for 3-5 minutes

3- Staining
   - Stain the smear with Giemsa or Wright's stain for 1-3 minutes respectively
   - Rinse the slide with distilled water at room temperature
   - Drain off the water and leave the slide to dry

4- Cover slipping
- Place a drop of Canada balsam on the smear and then mount the cover slip

5- Observation
- examine under dry objective or oil immersion
- count about 100-200 cells and take average

V. WBCs Staining Dyes

1- Wright's stain:

Preparation of stain
- Place 0.5gm of dry Wright's stain powder in a mortar
- Add 300ml of absolute methyl alcohol
- Grind the mixture and pour in dark bottle
- Shake the bottle each day for about two weeks, and then filter

Accelerated method
- Place 0.3gm of dry Wright's stain powder in a mortar and over lay with 3ml of glycerol, grind thoroughly
- Rinse with 100ml of absolute methyl alcohol and place in dark container
- Mix with magnetic stirrer for about 1 wk without heat
- Filter before use

2- Giemsa stain
- 1 g Giemsa stain powder is dissolved in 66 mL glycerol and heated to 56°C for 90 to120 minutes.
- After addition of 66 mL absolute methanol and thorough
- mixing, the solution is left at room temperature in a closed
- container.
- It must be filtered before use.

3- Leishmen's stain
- Grind 0.15gm of Leishmen's stain powder with small amounts of absolute methyl alcohol until an even suspension is obtained.
- A total of 100ml methanol is added to produce complete solution
- Pour in to a dark bottle and age for a few weeks prior to use

4- Wright's-Giemsa stain (modified Wright's stain) Preparation of stain

- 500mg of dry Wright's stain powder and 50mg of Giemsa stain powder are ground in a mortar with 100ml of absolute methyl alcohol (acetone free)
- Allow to stand for 24-48 hrs before using
- Keep well stoppered to prevent evaporation of alcohol or absorption of water vapor

VI. Hemoglobin Determination

- Method: Acid hematin method
- Requirements: Sahlis instrument, blood sample

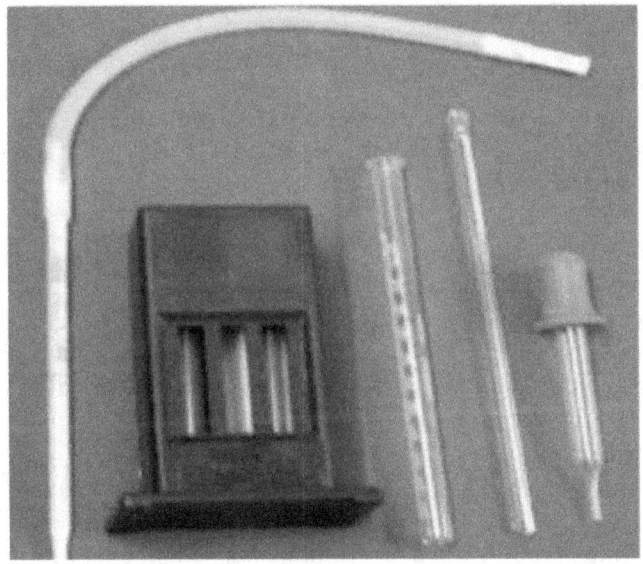

Procedure:

1. Take 0.1N HCl (1%) into central graduated tube up to mark 2.
2. Suck the blood exactly up to mark 20 (20 µl) with the help of sahlis pipette.
3. Transfer the blood from pipette to central graduated tube of the hemometer.
4. Mix it well with the help of stirrer or rod and allow it to react for two minute.
5. Make up with distilled water by adding drop by drop until the color matches with the Standard comparator tube and mix well.

6. When the color matches take out and record the values on the side as gm/100ml and or in percentage.
7. Repeat 5 to 6 times and take the average value

Hematocrit Determination (PCV)
Principle:
Hematocrit is the ratio of the total volume of RBC's to that of whole blood expressed as percentage (%) (Whole blood = total volume of cells + plasma). The second synonym for hematocrit is PCV (Packed Cell Volume). The procedure is easy to perform, whole blood is centrifuged in a narrow tube (capillary tube), cellular elements will be separated from the plasma, after centrifugation blood will be separated into 3 layers :
(1) Bottom layer contains packed RBC's,
(2) Middle layer contains WBC's and Platelets (on top of RBC's),
(3) Upper plasma layer.

Sample:
- EDTA anticoagulated whole venous blood
- Heparin or directly from a finger prick, to a heparin coated capillary tube.

Apparatus and Materials:
1-Microhematocrit centrifuge.
2-Modeling clay (seal material).
3-Capillary tubes (7 cm long, 1mm diameter)
4-Hematocrit measuring device reader or a conventional ruler.

Procedure
The blood is filled in to a micro hematocrit tube (3/4th) and seals it with sealer.

Centrifuge the filled hematocrit tube in a hematocrite centrifuge at 2000 rpm for 4-5 minutes.

Read the value (the tube) with hematocrit reader and record the result.

Clay end outside , away from the center

Balance the microcentrifuge

clay end outside away from the center

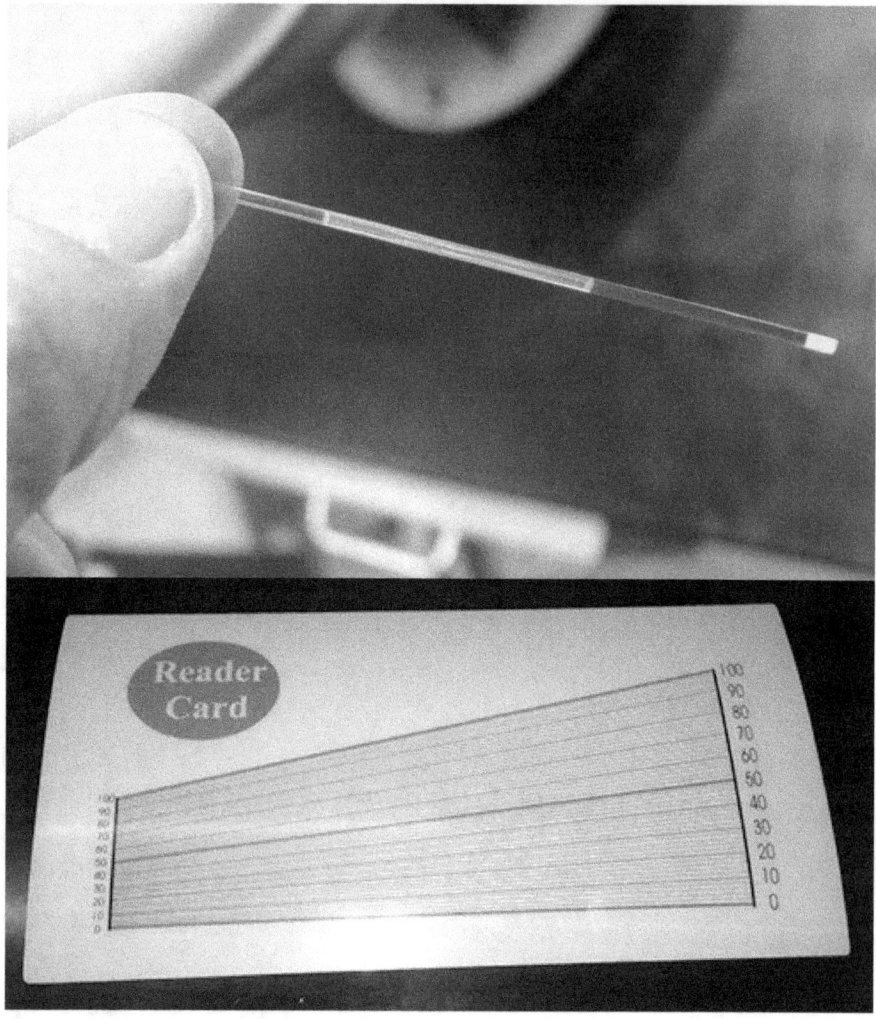

VII. Determination of Erythrocyte Sedimentation Rate (ESR)
Principle:

- It is measured by the degree of settling of red blood cells in a specific time period, usually one hour.
- There are several methods for determination of the ESR; the most common are the **Wintrobe** and the **Westergren**, named for the developers of the procedure.
- The erythrocyte sedimentation rate (ESR), also called a sedimentation rate or Biernacki Reaction, is the rate at which red blood cells precipitate in a period of 1 hour.
- The ESR is governed by the balance between pro sedimentation

factors, mainly fibrinogen, and those factors resisting sedimentation, namely the negative charge of the erythrocytes (zeta potential). When an inflammatory process is present, the high proportion of fibrinogen in the blood causes red blood cells to stick to each other. The red cells form stacks called 'rouleaux' which settle faster.

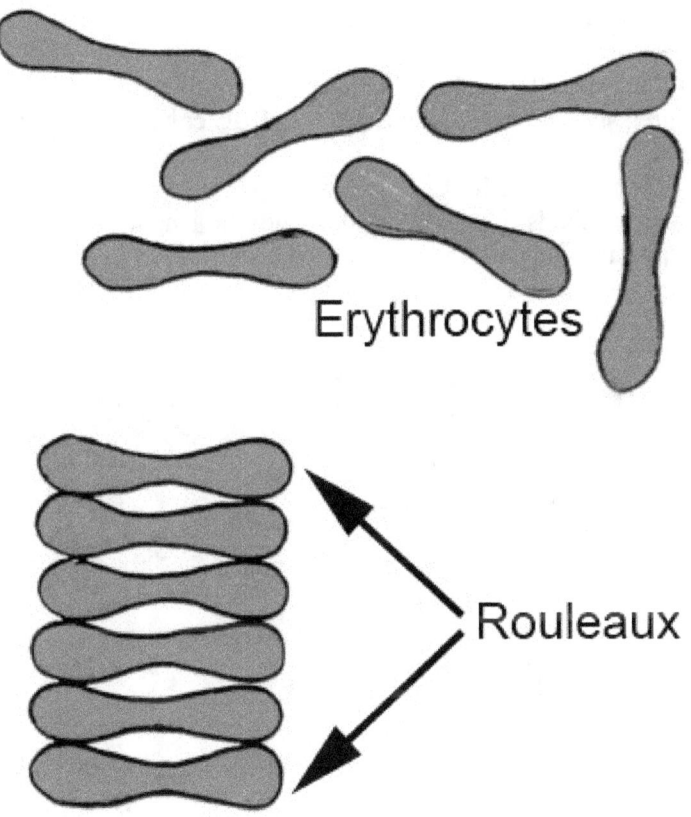

Erythrocytes

Rouleaux

A- Westergren Method

Westergren pipette is open at both the ends. It is 30 cm in length and 2.5 mm in diameter. The lower 20 cm are marked with 0 at top and 200 at bottom.

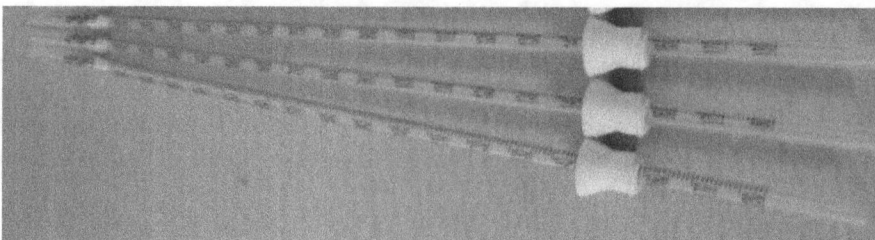

The anticoagulant used in this method is 3.8% tri-sodium citrate solution. 0.4 ml of tri-sodium citrate is added in 2 ml of blood.

Procedure:

Fill the pipette by sucking blood up to 0 marks and fix it vertically in Westergren stand. Read the upper level of RBC column exactly after 1 hr.

B- Wintrobe Method

Wintrobe tube is open at one side only. The length of Wintrobe tube is 11 cm and the diameter is 2.5 mm. The lower 10 cm are marked. The marking is 0 at top and 100 at bottom for ESR, and it is also used for **PCV**

.

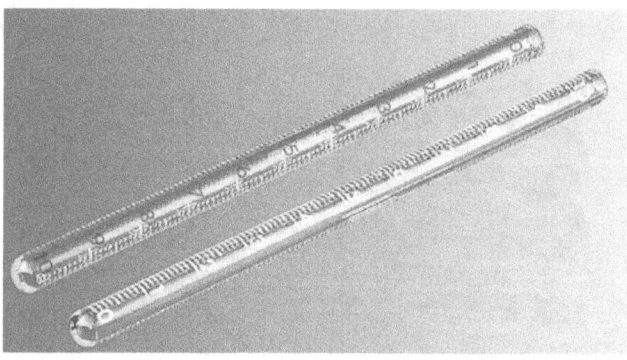

The anticoagulant used in Wintrobe Method is EDTA solution. 0.4 ml of anticoagulant is required for 2 ml of blood. (Packed Cell Volume).

Procedure:

With the help of long necked pasture pipette or a special syringe, fill the Wintrobe tube upto '0' mark. Place the tube in an exactly vertical position in a Wintrobe stand. Read the upper level of RBC column exactly after 1 hr.

VIII. Coagulation Time Determination (Whole Blood Clotting Time)

Procedure:
- A skin puncture is made and wiping away the first drop, fill a special capillary tube with blood noting the time when the blood first appeared
- Holding the tube between the thumb and index finger of both hands, gently break the tube every second until a strand of thread fibrin is seen extending across the gap between the ends of the tube
- The interval between the appearance of the blood and the appearance of the fibrin stand is the coagulation time

IX. Bleeding Time

Determination of bleeding time is a simple and sometimes useful tool for evaluating the efficiency of the capillary – platelet aspect of homeostasis

Method:

Dukes method

Protocol:
- Make a moderately small, deep puncture in clean, sterile blood lancet or sterile needle, and note the time when blood first appears
- Remove the drops of blood with filter paper every 30 second being careful not to touch the skin. The use of highly absorbent paper such as cleaning tissue tends to prolong bleeding time due to more effective removal of surface blood
- Note the end point, when blood no longer appears from the puncture site

X. Blood group and Rh-factor determination

The following table shows characteristics of the ABO blood group

Blood groups	Antigens (on blood cells)	Antibodies (in plasma/serum)
A	A	Anti - B
B	B	Anti - A
O	No antigen	Anti – A and anti – B
AB	A&B	No antibody

Method:
- Place one drop anti – A anti body (blue) on the first bore of the

plastic plate
- Place one drop anti – B anti body (yellow) on the second bore of the plastic plate
- Add a drop of blood in to each of the above anti-serums
- Using a separate clean stick for each bore , mix the blood and the antiserums
- Observe for agglutination reactions following few minutes
- Similar steps follow for Rh blood , except you use anti – D antibody

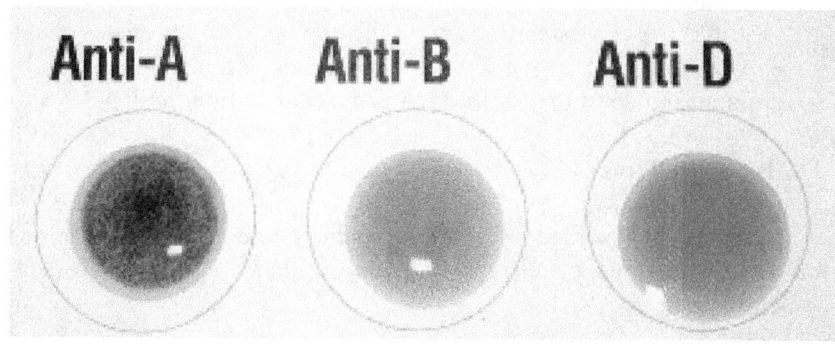

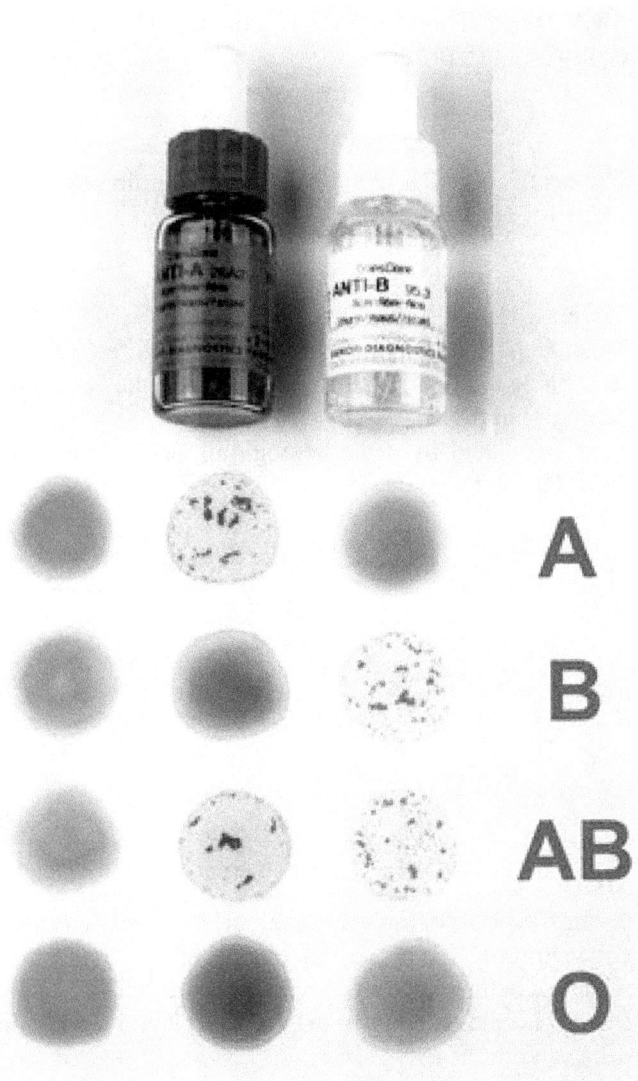

A

B

AB

O

Red Blood Cell Indices
1- **MCV**
- Mean Cell (Corpuscular) Volume, is the average volume of red cells.

- MCV is calculated from the hematocrit (HCT), and the Red Blood Cells Count (RBC count).

$$MCV = \frac{hematocrit(\%) \times 10}{RBC \; count(millions \, / \, mm^3 \; blood)}$$

- The results of MCV are expressed in femtoliters (fl).
- $1 \; fl = 1 \times 10^{-15} \; L$.

2- MCH

- Mean Cell Hemoglobin, is the hemoglobin content in the average red blood cell, or in other words, the average weight of hemoglobin per RBC.
- It is calculated from the hemoglobin concentration (Hb), and the total RBC count.

$$MCH = \frac{Hb \; (g/dL) \times 10}{RBC \; (\times 10^{12} / L)}$$

- Results of MCH are expressed in picograms (pg).
- $1 \; pg = 10^{-12} \; g$.

3- MCHC

- Mean Cell Hemoglobin Concentration, is the average hemoglobin concentration in 100 cc red blood cells. It indicates the average weightof hemoglobin as compared to the cell size.
- MCHC is calculated from the hematocrit and hemoglobin.

$$MCHC = \frac{Hb \; (g/dL) \times 100}{Hematocrit \; (\%)}$$

- Results of MCHC are expressed in percentage gm/dl.

4- RDW

- The red blood cell distribution width, or, is a measure of the variation of red blood cell (RBC) width that is reported as part of a standard complete blood count.
- The "width" in RDW is actually misleading, since it in fact is a

measure of deviation of the *volume* of RBCs, and not directly the diameter.

- Mathematically the RDW is calculated with the following formula:

RDW = (Standard deviation of MCV/ meanMCV) × 100

I. Estimation of Malonaldehyde using Ohkawa method

Principle:

In the TBA test reaction one molecule of MDA reacts with two molecules of TBA with the production of a pink pigment having an absorption maximum at 532-535 nm.

2-Thiobarbituric
Acid

MDA

The reaction should be performed at pH 2-3 at 90-100°.

Reagent preparation:

Reagent	Amount	Solvent
10% TCA	10 gm	100 ml H_2O
1% TBA	0.5 gm	50 ml hot H_2O

Procedures:

In dry test tubes add:

	Sample (ml)	Blank (ml)	Standard (For standard curve)
Sample	1	-	-
Standard	-	-	1
Deionized water	-	1	-
TCA	2	2	2
	The mixture was then centrifuged (3000 g) at room temperature for 10 min to separate proteins		
Supernatant	2	2	2
TBA	0.5	0.5	0.5
	Heated at 95°C on a water bath for 60 min using glass balls as condenser to generate the pink colored MDA		
	Cooling and read the absorbance of sample (A_{sample}) against blank at 532 nm.		

Calculation:

The molar extinction of MDA at 535nm is $= 0.156 \times 10^6$ Moles/litre/cm.

Malondialdehyde concentration (C):

$$C = \frac{Abs}{0.156 \times 10^6} \text{ moles/L}$$

$$C = \frac{Abs}{0.156 \times 10^9} \text{ moles/ml}$$

$$C = \frac{Abs}{0.156} \text{ nmoles/ml}$$

$$C_{plasma} = \frac{Abs}{0.156 \times ml_{plasma}} \text{ nmoles/ml}$$

$$C_{tissue} = \frac{Abs}{0.156 \times g.tissue \text{ used in homogenate}} \text{ nmoles/g.tissue}$$

You can use standard curve as follow:

Preparation of the MDA standard curve:

Reagent	Amount	Solvent
1 mM stock solution 1,1,3,3 tetraethoxypropane (TEP)	25 μL	100 ml H_2O
1% sulfuric acid	1 ml	99 ml H_2O (Add acid to water slowly)

Procedures:

Working standard was prepared by hydrolysis of **1 ml TEP** stock solution in **50 ml 1% sulfuric acid** and incubation for 2 hr at room temperature. The resulting MDA standard of 20 **nmol/ml**.

Preparation serial solution of MDA:

Concentration (nmol/ml)	Volume of MDA	Volume of 1%sulfuric acid
20	1ml	-
10.5	1ml	1.9
2.5	1ml	8
1.25	1ml	16
0.625	1ml	32

MDA standards should be used within 24 hours of preparation.

II. Estimation of Reduced glutathione (GSH) using Ellman method

Principle:

DTNB (Ellman's reagent) and glutathione (GSH) react to generate 2-nitro-5-thiobenzoic acid and GSSG. Since 2-nitro-5-thiobenzoic acid is a yellow colored product, GSH concentration can be determined by measuring absorbance at 412 nm.

2-Nitro-5-thiobenzoic acid

Reagent preparation:

Reagent	Amount	Solvent
0.3 M phosphate buffer (pH 7.4)		Chapter 1
0.1% EDTA	0.1 gm	100 ml H_2O With 6MNaOH (EDTA dissolve in alkaline water)
20% TCA	30 gm	150 ml H_2O
1 mM 5,5-dithiobis-2-nitrobenzoic acid (DTNB) M.Wt= 396.35	0.0594 gm	150ml 0.3 M phosphate buffer (pH 7.4)

Procedures:

In dry test tubes add:

Reagents	Sample (ml)	Blank
Sample	0.1	-
EDTA	0.9	-
TCA	1.5	-
	The mixture is allowed to stand for 5 min prior to centrifugation for 10 min at 3000 g rpm	
Supernatant	0.2	-
DTNB	1.8	1.8
	vortexes thoroughly	
	OD was read (**within 2-3 min** after the addition of DTNB) at 412 nm.	

Calculation:

$$GSH\ concentration = \frac{A\ x\ dilution\ factor}{13600\ x\ C}\ x\ 1000$$

$$dilution\ factor = \frac{5}{0.2}\ x\ \frac{2}{0.2} = 250$$

$$GSH\ concentration = \frac{A\ x\ 18.38}{C}$$

- **C**: ml for plasma used or mg protein for tissue used in homogenate.
- **$1.36 \times 104 \ M^{-1} \ cm^{-1}$**: extinction coefficient
- GSH concentration was expressed as mmoles /ml for plasma and as mmoles /mg protein for tissue.

III. Estimation of glutathione peroxidase using Rotruck method

Principle:

A known amount of enzyme was allowed to react with H_2O_2 in the presence of GSH for a specified time period. Then the remaining GSH was measured by the method of Ellman.

$$2GSH + H_2O_2 \xrightarrow{GPx} GSSG + 2H_2O$$

Reagent preparation:

Reagent	Amount	Solvent
Phosphate buffer, 0.4 M, pH 7	Chapter 1	
10 mM Sodium azide M.Wt 65.01	0.0065 gm	10 ml
0.4 mM EDTA M.Wt 292.24	0.0021 gm	20 ml alkaline H_2O
10% TCA	10 gm	100 ml H_2O
2 mM GSH Fresh M.Wt 307.32	0.012 gm	20 ml H_2O
0.2 mM H_2O_2 Fresh	Morality of 30 % H_2O_2 is 9.79 M $M1 \times V1 = M2 \times V2$ $9.79 \times V1 = (0.2 \times 10^{-3}) \times 1000$ ml **V1= 0.02ml**	999.98 ml H_2O
(0.1 mM) Ellman's reagent (5,50-dithiobis-2-nitrobenzoic acid)	0.071 gm	180ml of 0.3 M phosphate buffer (pH 7.4)

Procedures:

In dry test tubes add:

Reagents	Sample (ml)	Blank
phosphate buffer	0.2	0.2
EDTA	0.2	0.2
Sodium azide	0.1	0.1
Sample	0.5	-
GSH	0.2	0.2
H_2O_2	0.1	0.1
	This mixture and incubated at 37°C for **10 minutes**. The reaction was arrested by the addition of	
TCA	0.5	0.5
Supernatant	0.2	0.2
Ellman's reagent	1.8	1.8
	OD was read (within 2-3 min after the addition of DTNB) at 412 nm	

$$GPx \text{ } specific \text{ } activity$$
$$= \frac{(Ablank - sample) \text{ } x \text{ } dilution \text{ } factor}{13600 \text{ } x \text{ } C} \text{x} 1000$$

$$dilution \text{ } factor = \frac{5}{0.2} \text{ } x \text{ } \frac{2}{0.2} = 250$$

GPx specific activity
$$= \frac{(Ablank - sample) \text{ } x \text{ } 18.38}{C}$$

- C: ml for plasma used or mg protein for tissue used in homogenate
- Specific activity of GPx was expressed as mmoles/ml for plasma and as mmoles /mg protein for tissue.

IV. Estimation of glutathione–S– Transferase using Habig method

Principle:

GST catalyzes the conjugation of L-glutathione to 1-chloro-2,4-dinitrbenzene (CDNB) through the thiol group of the glutathione

$$GSH + CDNB \xrightarrow{GST} GS - CDNB \; congugate + HCl$$

The reaction product, GS-DNB Conjugate, absorbs at 340 nm. The rate of increase in the absorption is directly proportional to the GST activity in the sample.

Reagent preparation:

Reagent	Amount	Solvent
0.1M Sodium phosphate buffer pH 6.5	Chapter 1	
9.2 mM GSH (M.Wt 307.32)	$\dfrac{307.32 x (9.2 x 10^{-3}) x 20}{1000} =$ $0.057 \; gm$	20 ml H$_2$O
0.1 M CDNB M.Wt 202.55	$\dfrac{202.55 x 0.1 x 2}{1000} =$ $0.04 \; gm$	2 ml Ethanol (95%)

Procedures:

In dry test tubes add:

Reagent	Sample (ml)	Blank (ml)
Sodium phosphate buffer	1.5	1.5
GSH	0.2 ml	0.2 ml
CDNB	0.02 ml	0.02 ml
Deionized water	-	0.1
Sample	0.1	-
	(Start stopwatch) Record first absorbance	
The absorbance was measured at 340 nm and at the temperature 25⁰C spectrophotometrically. The increase in absorbance was recorded for a total 5 min.		

Calculation:

$GST\ activity\ (\mu mol/ml/min)$
$$= \frac{(\Delta A_{sample} - \Delta A_{BLANK}) x\ dilution\ factor}{9.6\ x\ ml_{sample}}$$

$GST\ activity$ (μmol/mg protein/min)

$$= \frac{\left(\Delta A_{sample} - \Delta A_{BLANK}\right) x\ dilution\ factor}{9.6\ x\ mg.tissue}$$

Dilution factor $= \frac{volume_{final}\ (1.82)}{volume_{initial}\ (0.1)} = 10.82$

- ΔA_{EXP}: Increase of the optical density of the sample per 5 min. $= \frac{A5-A1}{5}$
- ΔA_{EBLANK}: Increase of the optical density of the blank per 5 min. $= \frac{A5-\Delta1}{5}$

- **9.6 mM-1 $^{-1}$:** The extinction co-efficient of Glutathione-1-Chloro-2,4-Dinitrobenzene conjugate at 340 nm

V. Estimation of superoxide dismutase using Kakkar method

Principle:

SOD activity was measured based on the inhibition extent of amino blue tetrazolium formazan formation in the mixture of Nicotinamide Adenine Dinucleotide, Phenazine MethoSulphate and Nitroblue Tetrazolium

(NADH–PMS–NBT) system. The color formed at the end of the reaction can be extracted into butanol and measured at 560nm.

Reagent preparation:

Reagent	Amount	Solvent
0.052 M Sodium pyrophosphate decahydrate (pH 8.3) Molecular Weight 446.06	1.39 gm	80 ml H_2O. The pH was adjusted to 8.3 with 1N HCl, volume was made up to 100 ml with H_2O and stored at 4°C
186μm phenazine methosulphate (PMS) Mwt 306.34.	0.0014gm	25 ml H_2O
Nitroblue tetrazolium (NBT) (300μ m) M.Wt 817.6356	0.025 gm	100 ml H_2O and stored at 4°
780μM Nicotinamide Adenine Dinucleotide (NADH): M.Wt 709.40	14 mg	25 ml H_2O. The solution was prepared freshly each time

Procedures:

In dry test tubes add:

Regent	Sample ml	Plasma ml
Sample	0.1	0.05
Deionized water	-	1.35
Sodium pyrophosphate	1.2	
PMS	0.1	
NBT	0.3	
NADH	0.2 Start stopwatch	
	After incubation at 30 °C for 90 sec, the reaction was stopped by the addition	
glacial acetic acid	1	
n-butanol	4 mixture was stirred vigorously	
	The mixture was allowed to stand for 10 min, centrifuged and absorbance of the upper butanol layer recorded at 560 nm **against n-butanol (blank)**	

Calculation:

One unit of SOD was defined as that amount of enzyme that inhibits the rate of reactions by 50% under specified conditions.

$$\text{\% of inhibition} = \frac{A_{control} - A_{exp}}{A_{control}} \times 100$$

$$\text{Specific activity (U/g)} = \frac{\text{\% of inhibition} \times \text{dilution factor}}{50 \times C}$$

$$\text{Dilution factor} = \frac{4.1}{0.1} \quad \text{for tissue}$$

$$\text{Dilution factor} = \frac{4.05}{0.05} \quad \text{for plasma}$$

- C: ml for plasma used or g protein for tissue used in homogenate.

VI. Estimation of catalase using Aebi method

Kinetic method

The principle:

The principle was based on the hydrolyzation of H_2O_2 and decreasing absorbance at 240 nm.

$$2H_2O_2 \xrightarrow{catalase} 2H_2O + O_2$$

Reagents:

Reagent	Amount	Solvent
50mMphosphate buffer (pH 7.0)	Chapter 1	
30 mM H$_2$O$_2$ **Fresh prepare**	Mt = 34.01 % of solution=30, Density = 1.1 g/cm^3 $M = \dfrac{D \times \% \; solution \times 10}{M.Wt} = \dfrac{1.1 \times 30 \times 10}{34.01} = 9.79$ Morality of 30 % H$_2$O$_2$ is 9.79M M1 x V1= M2 xV2 9.79 x V1=(30 x 10^{-3}) x 100 ml **Take V1= 0.31ml H$_2$O$_2$**	Add 99.69 ml dist.H$_2$O

Procedures:

In dry test tubes add:

Reagent	volume (ml)
phosphate buffer	1.9
Sample	0.1
30 mM H_2O_2	1 Start stopwatch
	Record the initial and final absorbance in a one-minute period at **240nm**.

Calculation:

Specific activity (units/mg of protein/min) =

$$\frac{\Delta A \; x \; dilution \; factor \; 1000}{43.6 \; x \; mg \; protein}$$

$$= \frac{\Delta A \; X \; 30 \; x \; 1000}{43.6 \; x \; mg \; protein}$$

Specific activity (units/mg of protein/min)

$$= \frac{\Delta A \; X \; 688.073}{mg \; protein}$$

Where: $43.6 M^{-1}cm^{-1}$: molar extinction coefficient

VII. Estimation of Catalase using Sinha method

Colorimetric method

Principle:

Catalase first reacts with H_2O_2 to produce water and oxygen:

$$2H_2O_2 \xrightarrow{\text{catalase}} 2H_2O + O_2$$

The dichromate / acetic acid reagent can be thought of as a "stop bath" for catalase activity. The hydrogen peroxide which hasn't been split by the catalase will react with the dichromate to give a blue precipitate of perchromic acid. This unstable precipitate is then decomposed by heating to give the green solution which measure at 530 nm.

Reagent preparation:

Reagent	Amount	Solvent
0.01Mphosphate buffer, pH 7.0,	Chapter 1	
0.2 M H_2O_2	Morality of 30 % H_2O_2 is 9.79 M M1 x V1= M2 x V2 9.79 x V1=0.2x 100 ml **Take 2.04ml H_2O_2**	Add 97.96 ml Deionized water
5%potassium dichromate	5 gm	100 ml
potassium dichromate and glacial acetic acid mixed in a 1:3 ratio		150 ml acetic acid + 50 ml potassium dichromate

Procedures:

In dry test tubes add:

	Tissue (ml)	Control (ml)
phosphate buffer	1	1
Sample	0.1	-
H_2O_2	0.5 ml	0.5
	After 1 min. stop reaction by addition	
Dichromate acetic acid mixture	2 ml	2
	boiling water bath for 10 minutes, cooled and measure the color developed against air at 530 nm	

Calculation:

$$\Delta A = A_{control} - A_{sample}$$

Plot the H_2O_2 Standard Curve. Apply the ΔA to the H_2O_2 standard curve to get **C µmol** of H_2O_2 decomposed by catalase in 1min reaction.

$$catalse \text{ specific } activity = \frac{Cx\,16}{N}$$

- **N**: ml for plasma used or mg protein for tissue used in homogenate
- The activity of catalase was expressed as µmoles /min/ml plasma and as µmoles /min/mg protein.

Preparation of Standard Curve:

H_2O_2 (µM)	Take x volume of H_2O_2	Add y volume of H_2O	Total volume
160	≈16.34 µl from 30 % H_2O_2	≈999.9837 ml	1000 ml
130	1.63 ml of 160 µM H_2O_2	0.38ml	2 ml
100	1.54ml of 130 µM H_2O_2	0.46ml	2 ml
70	1.40 ml of 100 µM H_2O_2	0.60 ml	2 ml
40	1.14ml of 70 µM H_2O_2	0.86ml	2 ml
10	0.50ml of 140 µM H_2O_2	1.50ml	2 ml

VIII. Determination of nitric oxide using Montgomery method

Principle:

NO reacts with oxygen to produce stable products (nitrate and nitrite). Sulfanilic acid is quantitatively converted to a diazonium salt by reaction with nitrite in acid solution. The diazonium salt is then coupled to N-(1-naphthyl)ethylenediamine, forming an azo dye that can be spectrophotometrically quantitated based on its absorbance at 548 nm.

Sulfanilic acid   diazomium salt N-(1-naphthyl)ethylenediamine dihydrochloride

azo dye

Reagent preparation

Reagent	Amount	Solvent
0.1% **N-(1-naphthyl)ethylenediamine** **dihydrochloride**	0.1 gm	Dissolve 100 mL of H2O (may require heat + stirring) Store in dark at 4°C
5% phosphoric acid	5 gm	100 mL of H2O
1% Sulfanilic acid	1gm	100 ml phosphoric acid (5%)
300μMSodium Nitrite stock solution	0.0207gm	1000 ml deionized water

Preparation of the Griess Reagent:

N-(1-naphthyl)ethylenediamine	**5ml**
sulfanilic acid	5ml

Procedures:

In dry test tubes add:

Reagent	Sample	Blank
Griess Reagent	100 µL	100 µL
sample	300 µL	-
deionized water	2.6 mL	2.9 mL
	Incubate the mixture for 30 minutes at room temperature. Measure the absorbance of the nitrite-containing sample at 548 nm relative toblank	

Calculation:

Preparation standard curve

NaNO$_2$ concentration	Take x volume of NaNO$_2$ (ml)	Add y volume of deionized water (ml)
250µM	1.67 ml of 300µL	1.67
200µM	1.60 ml of 250µL	1.60
150µM	1.50 ml of 200µL	1.50
100µM	1.33 ml of 150µL	1.33
50µM	1.00 ml of 100µL	1.00
10 µM	0.4 ml of 50µL	1.6

- **Plot a standard curve of sodium nitrite (NaNO$_2$) concentration (x-axis) against absorbance (y-axis).**

- Convert absorbance readings to nitrite concentrations (μM/l)

REFERENCES

Aebi, H. (1984)Catalase in vitro. Methods enzymol 105, 121-126.

Caliezi C, Reber G, Lämmle B, de Moerloose P, Wuillemin WA. (2000) Agreement of D-dimer results measured by a rapid ELISA (VIDAS) before and after storage during 24 h or transportation of the original whole blood samples. Thromb Haemost; 83: 177-8.

Clark , W.M. & Lubs, H.A., 1917. J. Bacteriol., 2, p.1.

Dawson, R.M.C., Elliott, D.C., Elliott, W.H. & Jones, K., 1986. Data for Biochemical Research. Oxford: Clarendon Press.

Delory, G.E. & King, E.J., 1945. Biochem. J., 39, p.245.

Diehl, K.-H. et al. (2001). "A Good Practice Guide to the Administration of Substances and Removal of Blood, Including Routes and Volumes", J. Appl. Toxicol., 21, 15–23

Ellman, G.L., (1959). Tissue sulfhydryl groups. Arch. Biochem. Biophys., 82, 70–77.

Erwa W, Bauer FR, Etschwaiger R, Steiner V, Scott CS; Sedlmayr P. (1998)Analysis of aged samples with the Abott CD 400 hematology analyzer. Eur J Lab Med; 6: 4-15.

Gomori, G., 1946. Proc. Soc. Exp. Biol. Med., 62, p.33.

Good , N.E. & Izawa, S., 1972. Hydrogen ion buffers. Methods Enzymol., 24, pp.53-68.

Gottfried, S.P. and Gerard M.N. (1987). Determination of hematological parameters. Clinical Chemistry, 19,1077-1080.

Guder WG, da Fonseca-Wollheim F, Heil W, Müller-Plathe O, Töpfer G, Wisser H, et al. (1996). Wahl des optimalen Probenvolumens. Klin Chem Mitt; 27: 106-7.

Guide to the Care and Use of Experimental Animals, (1993). 2nd Edition Canadian Council on Animal Care, Canada

Habig, W.H., Pabst, M.J. and Jakoby, W.B. (1974). Glutathione S-transferases. The first enzymatic step in mercapturic acid formation. J. Biol. Chem. 249, 7130-7139.

Holmes, W., 1943. Anat. Rec., 86, p.163.

ISO GUIDE 30. Terms and definitions used in connection with reference materials. 1992; 30, 2nd ed.

Janet Hoff, LVT, RLATG(2000): Methods of Blood Collection in the Mouse. Lab Animal Technique. 29(10)

Kakkar, P. B. and Viswanathan , P. N. (1984) A modified spectrophotometric assay of superoxide dismutase. Indian J.l Biochem. Biophys. 21, 130-132.

Leonard PJ, Persaud J, Motwani R. (1971).The estimation of plasma albumin by BCG dye binding on the Technicon SMA 12/60. Clin Chim Acta; 35: 409.

Lillie, R.D., 1948. Histopathologic Technique. Philadelphia, PA: Blakiston.

McIlvaine, T.C., 1921. J. Biol. Chem., 49, p.183.

Mohan, C., 2006. Buffers A guide for the preparation and use. United States: EMD Biosciences.

Montgomery HAC, Dymock JF(1961). The determination of nitrate in water. Analyst; 86: 414-416.

Neumaier M, Braun A, Wagener C. (1998). Fundamentals of quality assessment of molecular amplification methods in clinical diagnostics. Clin Chem; 44: 12-26.

Ohkawa, H., Ohishi, N. & Yagi, K., (1979). Assay for lipid peroxides in animal tissues by thiobarbituric acid reaction. Anal.Biochem. 95, 351-358.

Parasuraman S, Raveendran R and Kesavan R. (2015): Blood sample collection in small laboratory animals. J.Pharmacol. Pharmacotherap.,1(2):87-93.

Perrin, D.D. & Dempsey, B., 1974. Buffers for pH and Metal Ion Control. London: Chapman & Hall.

Pilz J, Meineke I, Gleiter CH. (2000). Measurement of free and bound malondialdehyde in plasma by high-performance liquid chromatography as the 2,4-dinitrophenylhydrazine derivative. J Chromatogr B Biomed Sci Appl; 742: 315-25

Plumel, M., 1949. Bull. Soc. Chim. Biol., 30, p.129.

Rotruck JT, Pope AL, Ganther HE, Swanson. AB, Hafeman DG and *Hoekstra WG , (1973).* Selenium biochemical role as component of glutathione peroxidase*Science, 179, 588-590..*

Sambrook, J. & Russell, D.W., 2001. Molecular Cloning: A Laboratory Manual. New York: CSHL Press Cold Spring Harbor. p.A1.3.

Sinha, A. K. (1972).Colorimetric assay of catalase. Anal. Biochem. 47, 389-394.

Sorensen, S.P.L., 1909. Biochem. Z., 21, p.131.

UBC Animal Care Guidelines SOP: ACC-2012-Tech02

Walpole, G.S., 1914. J. Chem. Soc., 105, p.2501.

William & Wilkins, 1928. Determination of Hydrogen Ion. Baltimore.

Wolfensohn, S. and Lloyd, M. (1998). 2nd Edition, Blackwell Science Ltd. Guidelines for survival bleeding of mice and rats.

ABOUT THE AUTHOR

Dr. Ayman Saber Mohamed was born in Giza, Egypt, in 1984. He received the B. Sc.degree in chemistry and zoology from faculty of science, Cairo University, Egypt, in 2011. He joined the Department of zoology, Faculty of science, Cairo University as a Demonstrator in 2013. In 2014, he got the M. Sc.degree and became teacher assistant of molecular and integrated physiology till now.